Introduction to Soil Physics

Introduction to
Soil Physics

Natasha Kennedy

Larsen & Keller
www.larsen-keller.com

Introduction to Soil Physics
Natasha Kennedy
ISBN: 978-1-64172-097-7 (Hardback)

© 2019 Larsen & Keller

Larsen & Keller

Published by Larsen and Keller Education,
5 Penn Plaza,
19th Floor,
New York, NY 10001, USA

Cataloging-in-Publication Data

Introduction to soil physics / Natasha Kennedy.
    p. cm.
Includes bibliographical references and index.
ISBN 978-1-64172-097-7
1. Soil physics. 2. Soil science. 3. Agricultural physics. I. Kennedy, Natasha.
S592.3 .I58 2019
631.4--dc23

For more information regarding Larsen and Keller Education and its products, please visit the publisher's website www.larsen-keller.com

Table of Contents

Permissions

Index

Preface

The study of the physical properties and physical processes of the soil is under the domain of soil physics. It is a multidisciplinary science that integrates the principles of physics, physical chemistry, meteorology and engineering for the study of soil components, their phases and dynamics. Such investigations drive solutions to problems in agriculture, ecology and engineering. This textbook is a compilation of chapters that discuss the most vital concepts in the field of soil physics. Different approaches, evaluations and methodologies have been included in this book. As this field is emerging at a rapid pace, the contents of this book will help the readers understand the modern concepts and applications of the subject.

Given below is the chapter wise description of the book:

Chapter 1, The study of the soil, its physical properties and processes is under the domain of soil physics. Such studies are crucial for addressing the problems in agriculture and ecology. This chapter has been carefully written to provide an introduction to soil and soil physics. **Chapter 2**, Soil acts as a habitat for soil organisms, a regulator of water quality, a recycling system for organic wastes and nutrients, a modifier of the composition of the atmosphere, an important provider of ecosystem services and a medium that supports plant growth. These services depend on the physical properties. The diverse physical properties of the soil such as texture, structure, density, porosity, color, resistivity, etc. and their significance have been extensively discussed in this chapter. **Chapter 3**, The vadose zone is the unsaturated part of the subsurface of the Earth between the surface of the Earth and the water table. The aim of this chapter is to explore the varied aspects of vadose zone, such as movement of water in unsaturated and saturated zone, soil-water content, soil-water content measurement methods, heat flux through soil, etc. for an in-depth understanding. **Chapter 4**, The rate of evaporation from the soil surface is influenced by various soil characteristics, environmental interactions and tillage. This chapter discusses in detail the process of evaporation through the soil, and includes the topics relevant to this study such as soil water redistribution during evaporation, vapor flow through soil and soil transpiration. **Chapter 5**, In order to study the phenomenon of solute transport in the soil, it is necessary to understand the processes by which nutrients are transported in the soil. This chapter has been written to provide a comprehensive understanding of this area of study. **Chapter 6**, Various gases are found in the air space that exists between soil components. These include carbon dioxide, nitrogen, oxygen, etc. This chapter closely examines the air flow through soil and includes various topics such as soil gas, soil aeration, air permeability, etc.

Indeed, my job was extremely crucial and challenging as I had to ensure that every chapter is informative and structured in a student-friendly manner. I am thankful for the support provided by my family and colleagues during the completion of this book.

Natasha Kennedy

Understanding Soil Physics

The study of the soil, its physical properties and processes is under the domain of soil physics. Such studies are crucial for addressing the problems in agriculture and ecology. This chapter has been carefully written to provide an introduction to soil and soil physics.

Soil

Soil is the mixture of minerals, organic matter, liquids, and gases covering most of the Earth's land surface and that serves, or has the ability to serve, as a medium for the growth of land plants. Although it may be covered by shallow water, if the water is too deep to support land plants (typically more than 2.5 meters), then the rock-covering mixture is not considered to be soil.

Loess field in Germany

Soil is vitally important to all life on land. It supports rooted plants, provides a habitat and shelter to many animals, and it is the home to bacteria, fungi, and other microorganisms that recycle organic material for reuse by plants.

While the general concept of soil is well established, the definition of soil varies, according to the perspective of the discipline or occupation using soil as a resource.

Soil is among our most important natural resources because of its position in the landscape and its dynamic, physical, chemical, and biologic functions. It has been both used and misused. On the positive side, human creativity is expressed in using soil for agriculture, gardening and landscaping, utilizing peat as an energy source, producing fertilizers to replenish lost nutrients, employing soils as building materials (such as adobe and mud brick), and transforming clay into eating and drinking vessels, storage containers, and works of art. On the other hand, anthropogenic activities have included fostering soil erosion and desertification through clear-cutting and overgrazing livestock, and contaminating soils by the dumping of industrial or household wastes.

The understanding of soil is incomplete. Despite the duration of humanity's dependence on and curiosity about soil, exploring the diversity and dynamic of this resource continues to yield fresh discoveries and insights. New avenues of soil research are compelled by our need to understand soil in the context of climate change and carbon sequestration. Our interest in maintaining the planet's biodiversity and in exploring past cultures has also stimulated renewed interest in achieving a more refined understanding of soil.

The earth's soil in general is sometimes referred to as comprising the pedosphere, which is positioned at the interface of the lithosphere with the biosphere, atmosphere, and hydrosphere. The scientific study of soil is called pedology or edaphology. Pedology is the study of soil in its natural setting, while edaphology is the study of soil in relation to soil-dependent uses.

Soil Components

Mineral Material

The majority of material in most soil is mineral. This consists of small grains broken off from the underlying rock or sometimes transported in from other areas by the action of water and wind. Larger mineral particles called sand and smaller particles called silt are the product of physical weathering, while even smaller particles called clay (a group of hydrous aluminium phyllosilicate minerals typically less than 2 micrometers in diameter) is generally the product of chemical weathering of silicate-bearing rocks. Clays are distinguished from other small particles present in soils such as silt by their small size, flake or layered shape, affinity for water and tendency toward high plasticity.

The mineral part of soil slowly releases nutrients that are needed by plants, such as potassium, calcium, and magnesium. Recently formed soil, for instance that formed from lava recently released from a volcano, is richer in nutrients and so is more fertile.

Organic Material

As plants and animals die and decay they return organic (carbon-bearing) material to the soil. Organic material tends to loosen up the soil and make it more productive for plant growth. Microorganisms, such as bacteria, fungi, and protists feed on the organic material and in the process release nutrients that can be reused by plants. The micro-organisms themselves can form a significant part of the soil.

Water and Air

Soil almost always contains water and air in the spaces between the mineral and organic particles. Most soil organisms thrive best when the soil contains about equal volumes of water and air.

Soil Classification

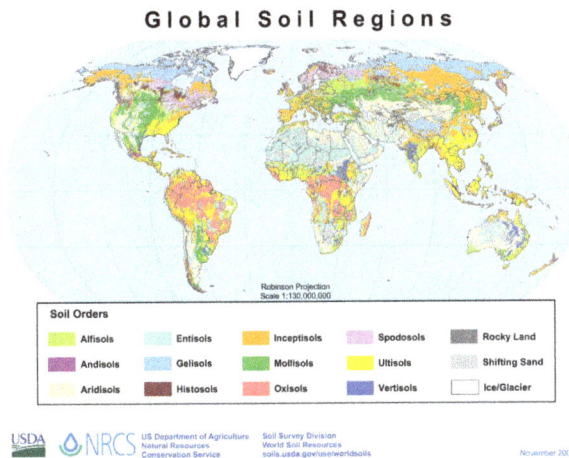

Map of global soil regions from the USDA

The World Reference Base for Soil Resources (WRB) is the international standard soil classification system. Development of this system was coordinated by the International Soil Reference and Information Centre (ISRIC) and sponsored by the International Union of Soil Sciences (IUSS) and the Food and Agriculture Organization (FAO) via its Land and Water Development division. It replaces the previous FAO soil classification system.

The WRB borrows from modern soil classification concepts, including United States Department of Agriculture (USDA) soil taxonomy. The classification is based mainly on soil morphology as an expression of pedogenesis, the creation of soil. A major difference with USDA soil taxonomy is that soil climate is not part of the system, except in so far as climate influences soil profile characteristics.

The WRB structure is either nominal, giving unique names to soils or landscapes, or descriptive, naming soils by their characteristics such as red, hot, fat, or sandy. Soils are distinguished by obvious characteristics, such as physical appearance (e.g., color, texture, landscape position), performance (e.g., production capability, flooding), and accompanying vegetation. A vernacular distinction familiar to many is classifying texture as heavy or light. Light soils have lower clay content than heavy soils. They often drain better and dry out sooner, giving them a lighter color. Lighter soils, with their lower moisture content and better structure, take less effort to turn and cultivate. Contrary to popular belief light soils do not weigh less than heavy soils on an air dry basis nor do they have more porosity.

Soil Characteristics

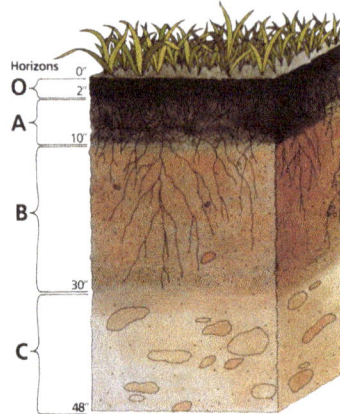

Soil horizons are formed by combined biological, chemical and physical alterations.

Soils tend to develop an individualistic pattern of horizontal zonation under the influence of site specific soil-forming factors. Soil color, soil structure, and soil texture are especially important components of soil morphology.

Soil color is the first impression one has when viewing soil. Striking colors and contrasting patterns are especially memorable. The Red River of the United States carries sediment eroded from extensive reddish soils like Port Silt Loam in Oklahoma. The Yellow River in China carries yellow sediment from eroding loessal soils. Mollisols in the Great Plains are darkened and enriched by organic matter. Podsols in boreal forests have highly contrasting layers due to acidity and leaching.

Soil color is primarily influenced by soil mineralogy. The extensive and various iron minerals in soil are responsible for an array of soil pigmentation. Color development and distribution of color within a soil profile result from chemical weathering, especially redox reactions. As the primary minerals in soil-parent material weather, the elements combine into new and colorful compounds. Iron forms secondary minerals with a yellow or red color; organic matter decomposes into black and brown compounds; and manganese forms black mineral deposits. These pigments give soil its various colors and patterns and are further affected by environmental factors. Aerobic conditions produce uniform or gradual color changes, while reducing environments result in disrupted color flow with complex, mottled patterns and points of color concentration.

Soil structure is the arrangement of soil particles into aggregates. These may have various shapes, sizes and degrees of development or expression. Soil structure influences aeration, water movement, erosion resistance, and root penetration. Observing structure gives clues to texture, chemical and mineralogical conditions, organic content, biological activity, and past use, or abuse.

Surface soil structure is the primary component of tilth. Where soil mineral particles are both separated and bridged by organic-matter-breakdown products and soil-biota exudates, it makes the soil easy to work. Cultivation, earthworms, frost action, and rodents mix the soil. This activity decreases the size of the peds to form a granular (or crumb) structure. This structure allows for good porosity and easy movement of air and water. The combination of ease in tillage, good moisture and air-handling capabilities, good structure for planting and germination are definitive of good tilth.

Soil texture refers to sand, silt and clay composition in combination with gravel and larger-material content. Clay content is particularly influential on soil behavior due to a high retention capacity for nutrients and water. Due to superior aggregation, clay soils resist wind and water erosion better than silty and sandy soils. In medium-textured soils, clay can tend to move downward through the soil profile to accumulate as illuvium in the subsoil. The lighter-textured, surface soils are more responsive to management inputs, but also more vulnerable to erosion and contamination.

Texture influences many physical aspects of soil behavior. Available water capacity increases with silt and, more importantly, clay content. Nutrient-retention capacity tends to follow the same relationship. Plant growth, and many uses which rely on soil, tends to favor medium-textured soils, such as loam and sandy loam. A balance in air and water-handling characteristics within medium-textured soils are largely responsible for this.

Soil and Its Environment

Soil and Plants

Soil is necessary for almost all land plants to survive and grow. A sprouting seed sends into the soil roots, which absorb water and dissolved minerals that the new plant needs for its growth. As the plant grows its root system expands though the soil and serves to support it. The soil stores water from rain and snow, allowing plants to have a continuous supply and helping to prevent destructive flooding. As plants lose their leaves, and when the plants themselves die, organic material is returned to the soil, which builds up and enriches the soil. In many environments, plants also protect the soil from erosion by holding it in place with their roots and sheltering it from the effects of wind and rain.

Soil and Animals

Since all animals depend on plants for their food, directly or indirectly, all land animals depend on the soil that makes plant life on land possible. Some animals, such as earthworms and moles, live all their lives in the soil itself, while others, such as ground squirrels and most ants, live in homes dug into the soil, protecting them from predators and from heat and cold and bad weather. Other animals, such as pigs, dig into the soil to find food, and others, such as beavers and some birds, use soil as a building material. Bison, elephants, and some other large animals cover themselves with soil for protection against sun and insects.

The wastes of animals enrich the soil and their digging mixes and loosens it; both of these activities benefit the further growth of plants. In New York State (United States), it is estimated the woodchucks turn over 1.6 million tons (1.63 million metric tons) of soil each year.

Soil in Nature

Soil formation processes never stop and soil is always changing. The long periods over which change occurs and the multiple influences of change mean that simple soils are rare. While soil can achieve relative stability in properties for extended periods of time, the soil life cycle ultimately ends in soil conditions that leave it vulnerable to erosion. Little of the soil continuum of the earth is older than Tertiary and most are no older than Pleistocene (Hole and McCracken 1973). Despite the inevitability of soil retrogression and degradation, most soil cycles are long and productive. How the soil "life" cycle proceeds is influenced by at least five classic soil forming factors: regional climate, biotic potential, topography, parent material, and the passage of time.

An example of soil development from bare rock occurs on recent lava flows in warm regions under heavy and very frequent rainfall. In such climates, plants become established very quickly on basaltic lava, even though there is very little organic material. The plants are supported by the porous rock becoming filled with nutrient-bearing water, for example, carrying dissolved bird droppings or guano. The developing plant roots themselves gradually breaks up the porous lava and organic matter soon accumulates but, even before it does, the predominantly porous broken lava in which the plant roots grow can be considered a soil.

Most of our knowledge of soil in nature comes from soil survey efforts. Soil survey, or soil mapping, is the process of determining the soil types or other properties of the soil cover over a landscape, and mapping them for others to understand and use. It relies heavily on distinguishing the individual influences of the five classic soil forming factors. This effort draws upon geomorphology, physical geography, and analysis of vegetation and land-use patterns. Primary data for the soil survey are acquired by field sampling and supported by remote sensing.

Geologists have a particular interest in the patterns of soil on the surface of the earth. Soil texture, color and chemistry often reflect the underlying geologic parent material and soil types often change at geologic unit boundaries. Geologists classify surface soils using the 1938 USDA soil taxonomy, but use the current version of USDA soil taxonomy to classify the buried soils that make up the paleopedological record. Buried paleosols mark previous land surfaces and record climatic conditions from previous eras.

Geologists use this paleopedological record to understand the ecological relationships in past ecosystems. According to the theory of biorhexistasy, prolonged conditions conducive to forming deep, weathered soils result in increasing ocean salinity and the formation of limestone.

Sample of an aerial photo from a published soil survey

Geologists and pedologists use soil profile features to establish the duration of surface stability in the context of geologic faults or slope stability. An offset subsoil horizon indicates rupture during soil formation and the degree of subsequent subsoil formation is relied upon to establish time since rupture.

Soil examined in shovel test pits is used by archaeologists for relative dating based on stratigraphy (as opposed to absolute dating). What is considered most typical is to use soil profile features to determine the maximum reasonable pit depth than needs to be examined for archaeological evidence in the interest of cultural resources management.

Soils altered or formed by man (anthropic and anthropogenic soils) are also of interest to archaeologists. An example is *Terra preta do Indio,* found in the Amazon river basin.

Soil Uses

A homeowner tests soil to apply only the nutrients needed.
Farmers practice the same testing procedure.

Due to their thermal mass, rammed earth walls fit in with environmental sustainability aspirations.

Gardening and landscaping provide common and popular experience with soils. Home-owners and farmers alike test soils to determine how they can be maintained and improved. Plant nutrients such as nitrogen, phosphorus, and potassium are tested for in soils. If a specific soil is deficient in these substances, fertilizers may provide them. Extensive academic research is performed in an effort to expand the understanding of agricultural soil science.

Soil has long been used as a building material. Soil-based wall construction materials include adobe, chirpici, cob, mudbrick, rammed earth, and sod. These materials often have the advantage of storing heat and protecting the interior of the building against extremes of heat and cold, while saving energy needed to heat and cool the building.

Organic soils, especially peat, serve as a significant fuel resource. Peat is an accumulation of partially decayed vegetation matter and forms in many wetlands around the world; approximately 60 percent of the world's wetlands are peat. The majority of peatlands are found in high latitudes. Peatlands cover around 3 percent of the global land mass, or about 4,000,000 km² (1,500,000 square miles). Peat is available in considerable quantities in Scandinavia: some estimates put the amount of peat in Finland alone to be twice the size of North Sea oil reserves. Peat is used to produce both heat and electricity, often mixed with wood. Peat accounts for 6.2 percent of Finland's yearly energy production, second only to Ireland. Peat is arguably a slowly renewable biofuel, but is more commonly classified as a fossil fuel.

Clay is another material taken from the soil that has been very important to humans, being used for eating and drinking vessels, storage containers, for works of art, and for other uses since prehistoric times.

Waste management often has a soil component. Using compost and vermicompost are popular methods for diverting household waste to build soil fertility and tilth. (Untreated human waste should not be used to improve soil in the case of agriculture intended for human consumption, because of the potential to spread parasites and disease.) The technique for creating *terra prêta do índio* in the Amazon basin appears to have started from knowledge of soil first gained at a household level of waste management. Industrial waste management similarly relies on soil improvement to utilize waste treatment products. Compost and anaerobic digestate (also termed biosolids) are used to benefit the soils of land remediation projects, forestry, agriculture, and for landfill cover. These products increase soil organic content, provide nutrients, enhance microbial activity, improve soil ability to retain moisture, and have the potential to perform a role in carbon sequestration.

Septic drain fields treat septic tank effluent using aerobic soil processes to degrade putrescible components. Pathogenic organisms vulnerable to predation in an aerobic soil environment are eliminated. Clay particles act like electrostatic filters to detain viruses in the soil adding a further layer of protection. Soil is also relied on for chemically binding and retaining phosphorus. Where soil limitations preclude the use of a septic drain field, the soil treatment component is replaced by some combination of mechanical aeration, chemical oxidation, ultraviolet light disinfection, replaceable phosphorus retention media and/or filtration.

For industrial wastewater treatment, land application is a preferred treatment approach when oxygen demanding (putrescible) constituents and nutrients are the treatment targets. Aerobic soil processes degrade oxygen demanding components. Plant uptake and removal through grazing or harvest perform nutrient removal. Soil processes have limited treatment capacity for treating metal and salt components of waste.

It has been suggested that building up the organic material in soils will have the effect of removing carbon from the atmosphere thereby helping to reverse or slow down any process of global warming, while at the same time increasing the soils' fertility.

Soil and Land Degradation

Light colored soils in northeast Iowa have lost their topsoil.
These soils are highly erodible and very steep.

Land degradation is a human induced or natural process that impairs the capacity of land to function. Soils are the critical component in land degradation when it involves acidification, contamination, desertification, erosion, or salination.

While soil acidification of alkaline soils is beneficial, it degrades land when soil acidity lowers crop productivity and increases soil vulnerability to contamination and erosion. Soils are often initially acid because their parent materials were acid and initially low in the basic cations (calcium, magnesium, potassium, and sodium). Acidification occurs when these elements are removed from the soil profile by normal rainfall or the harvesting of crops. Soil acidification is accelerated by the use of acid-forming nitrogenous fertilizers and by the effects of acid precipitation.

Soil contamination at low levels are often within soil capacity to treat and assimilate. Many waste treatment processes rely on this treatment capacity. Exceeding treatment capacity can damage soil biota and limit soil function. Derelict soils occur where industrial contamination or other development activity damages the soil to such a degree that the land cannot be used safely or productively. Remediation of derelict soil uses principles of geology, physics, chemistry, and biology to degrade, attenuate, isolate, or remove soil contaminants and to restore soil functions and values. Techniques include leaching, air sparging, chemical amendments, phytoremediation, bioremediation, and natural attenuation.

In the Pantanal, the world's largest wetland, damming and diking to separate the land from the water also prevents the natural flooding that replenishes the nutrients in the soil. This then requires greater amounts of fertilizers, which then tend to contaminate the surrounding ecosystem.

Desertification is an environmental process of ecosystem degradation in arid and semi-arid regions, or as a result of human activity. It is a common misconception that droughts cause desertification. Droughts are common in arid and semiarid lands. Well-managed lands can recover from drought when the rains return. Soil management tools include maintaining soil nutrient and organic matter levels, reduced tillage, and increased cover. These help to control erosion and maintain productivity during periods when moisture is available. Continued land abuse during droughts, however, increases land degradation. Increased population and livestock pressure on marginal lands accelerates desertification.

Soil erosional loss is caused by wind, water, ice, and movement in response to gravity. Although the processes may be simultaneous, erosion is distinguished from weathering, which occurs in situ, or "without movement," while erosion involves movement. Erosion is an intrinsic natural process, but in many places it is increased by human land use. Poor land use practices include deforestation, overgrazing, and improper construction activity. Improved management can limit erosion using techniques like limiting disturbance during construction, avoiding construction during erosion prone periods, intercepting runoff, terrace-building, use of erosion suppressing cover materials and planting trees or other soil binding plants.

Sediment in the Yellow River.

A serious and long-running water erosion problem is in China, on the middle reaches of the Yellow River and the upper reaches of the Yangtze River. From the Yellow River, over 1.6 billion tons of sediment flow each year into the ocean. The sediment originates primarily from water erosion in the Loess Plateau region of northwest China. The Taquiri River in the Pantanal area of Brazil is another classic site of erosion, leading to significant channel alteration—to the extent of the loss of over one hundred farms, branching of the river to where the channel is 30 percent of its former size, and loss of the fishing industry.

One of the main causes of soil erosion in is slash and burn treatment of tropical forests.

Soil piping is a particular form of soil erosion that occurs below the soil surface. It is associated with levee and dam failure as well as sink hole formation. Turbulent flow removes soil starting from the mouth of the seep flow and subsoil erosion advances upgradient.

Soil salination is the accumulation of free salts to such an extent that it leads to degradation of soils and vegetation. Consequences include corrosion damage, reduced plant growth, erosion due to loss of plant cover and soil structure, and water quality problems due to sedimentation. Salination occurs due to a combination of natural and human caused processes. Aridic conditions favor salt accumulation. This is especially apparent when soil parent material is saline. Irrigation of arid lands is especially problematic. All irrigation water has some level of salinity. Irrigation, especially when it involves leakage from canals, often raise the underlying water table. Rapid salination occurs when the land surface is within the capillary fringe of saline groundwater.

An example of soil salination occurred in Egypt in the 1970s after the Aswan High Dam was built. The source water was saline. The seasonal change in the level of ground water before the construction had enabled salt flushing, but lack of drainage resulted in the accumulation of salts in the groundwater. The dam supported irrigation, which raised the water table. A stable, shallow water table allowed capillary transport and evaporative enrichment of salts at the soil surface, depressing crop productivity below pre-project levels.

Preventing soil salination involves flushing with higher levels of applied water in combination with tile drainage.

Soil Physics

Soil physics is a relatively young science, which measures, predicts and controls the physical processes taking place in and through the soil. It uses theories, models, technologies and instruments, and benefits from laboratory and field experiments for a perpetual evolution and better fulfilling the future requirements. That soil physics followed two paths during its development. The first is the empirical one characterized by countless numbers of laboratory and field experiments where the results were only qualitatively interpreted. The second path is characterized by the real application of physical theories to the solution of transport processes in soil, to tillage and compaction of soil and many other processes and phenomena. By providing a description of these processes, soil physics constitutes an instrument essential for a sustainable management of soils, not only as a bridge for improving plant production, but also with regard to their protection and conservation. However, until recently, the knowledge of the soil physical system as a whole was recognized to be still fragmentary, because of the space and time variation of soil properties.

Soil physics: Development in Measurement Techniques

Development in instrumentation helped to explain simultaneously processes, satisfy theories and orient management decisions in agriculture (irrigation, drainage), environment (waste management and disposal, erosion, salinity control, etc.), and industry (remediation and clean-up). Mainly soil moisture, hydraulic conductivity and soil structure. In fact, remarkable progress has been made on instruments for measuring most soil physical properties (particle-size distribution (PSD), gas diffusivity, thermal conductivity, salinity, alkalinity, etc.). For all soil scientists, PSD constitutes a fundamental soil property correlated to most other soil properties. PSD can be measured either with simple apparatus (hydrometer and pipette) or with advanced equipment (sedigraph, laser, gamma- or X-ray). These instrumental developments are due to needed precision in data collection for accurate interpretation of space and time variation of soil properties. The satisfaction of these goals prompted the emergence of advanced theories and comprehensive mechanisms of most natural processes, the development of new mathematical tools (modeling and computer simulation, fractals, geostatistics, transformations), the creation of high precision instrumentation (computer assisted, less time constraint, increased number of measured parameters) and the scale sharpening of physical measurements, which range, from the micro-scale to the watershed).

Knowledge of the soil water status is of primary importance in soil physics. The standard method for measuring water content is the gravimetric method. This method becomes diffcult when measurements are required in dry climates because significant water losses can occur rapidly during handling and weighing. In the field, the neutron probe provides a non-destructive method for monitoring water content in the subsoil. Its main disadvantages are the high initial cost of the instrument, low degree of spatial resolution, diffculty of measuring moisture in the soil surface zone and the health hazard associated with exposure to neutron radiation. In recent years, electromagnetic or dielectric methods have been used increasingly. These include time-domain reflectometry or TDR. TDR has been applied to measure soil water content and/or salt concentration directly in the field through a calibration curve relating the dielectric constant of the soil material as a function of water content and/or salt concentration. This instrument is still a subject of research and development. However, other measurement techniques have not achieved much technical improvement and are used largely as originally developed. This is the case for matric potential, a critical variable in water management and agriculture, measured with tensiometer and porous pressure plate. Hydrodynamic properties (conductivity, diffusivity, sorptivity) are parameters of water flow theories. Hence, field measurements were required to validate these theories. Various instruments and devices were designed accordingly. Among recently developed equipment, Guelph, disk permeameter and tension infiltrometer are the most commonly used.

Mechanical analysis of soil structure can produce useful information for agronomical soil classification and guidelines for soil management. However, quantification of stable aggregates in soils is still largely dependant at early-developed method. Even though soil structure and aggregation are important parameters and are processes occurring under continuous changes in the environment, their measurement has been modified only slightly. The dry method developed by Chepil consisted of separation of aggregates in a rotary sieve with several sizes. In the high portion of the aggregate diameter (>0.83 mm), it is less likely that wind erosion will occur. The wet method developed by Kemper and Chepil was modified in 1986 by Kemper and Rosenau. It was meant to reflect the occurrence of water erosion. In order to quantify more precisely the soil structure, several attempts were made in the application of image analysis or using the computed tomography method. This latter related porosity and pore size distribution to soil management. Other methods exist, and the treatment consists in wetting the samples in liquids other than distilled water, such as ethanol.

Relationship to Other Disciplines

One would not study soil physics without carrying out frequent incursions into the other related disciplines (agronomy, hydrology, geology, mathematics, pedology, physics, sedimentology, etc.). For successful treatment of all classical and new environmental problems, soil physics completes its traditional theories with recently developed approaches in applied mathematics and instrumentation physics and its interaction with other scientific branches. Especially, hydrologists are concerned with water regime and water flow through the soil, while soil physics can explain the greenhouse effect that interests climatologists. Additionally, the environmental industry has contributed to a better understanding of many facets of soil physics. In other words, research demand in soil physics has increased considerably to satisfy specific and environmental problems (contamination of water resources, global warming, etc.).

In soil physics the primary interest is in mixtures of a solid phase, a liquid phase and a gaseous phase. Each of these phases is itself a mixture. The solid phase is a mixture of numerous inorganic and organic constituents, many of which are positively or negatively charged. The composition of the gaseous phase may more or less deviate from that of the atmosphere above the soil surface. The aqueous phase is a dilute aqueous solution, balancing the charges of the solid phase and being strongly buffered by the solid and gaseous phases. For the roots of plants the soil is a resilient provider of nutrients and receiver of excrements. For many purposes it is useful to regard the root system itself as an additional phase of the solid- liquid-gas mixture.

Theory of Soil Physics

1. Solid phase: Soil material less than 2 mm size constitutes the soil sample. It is broadly composed of inorganic and organic constitutes. Soils having more than 20% of org. constitutes are arbitrarily designated organic soils. Where inorganic constituents dominate, they are called mineral soils. The majority of the soils of India are mineral soils. It accounts for nearly 50% of the total volume and 95% without of the solid phase is made up of inorganic or mineral matter. The remaining 5% weight comprises of OM which is mainly derived from dead parts of the vegetation an animals. In inorganic constituents consist of silicates, certain preparation of carbonates, soluble salts, an free oxides of iron, aluminium and silicon. The humus and humus like fractions of the solid phase constitute the soil organic matter. Soil is the habitat for enormous number of living organisms like roots of higher plants (Soil Macro flora), bacteria, fungi, actinomycetes and algae (Soil Micro flora). A gram of fertile soil contains billions of these micro-organisms. The live weight of the micro-organisms may be about 4000 kg/ha may constitute about 0.01 to 0.4% of the total soil mass. Soil also consists of protozoa and nematodes (Soil Micro Fauna).

2. Liquid phase: About 50% of the bulk volume of the soil body is generally occupied by voids or soil pores which may be completely or partially filled with water. A considerable part of the rain which falls on soil is absorbed by the soil and stored in it to be returned to the atmosphere by direct evaporation or by transpiration through plants. The soil acts as the reservoir for supplying water to plants for their growth. The soil water keeps salts in solution which act as plant nutrients. Thus, the liquid phase is an aqueous solution of salts, when water drains from soil pores are filled with air.

3. Gaseous phase: The air filled pores constitutes the gaseous phase of soil system and dependent on that of the liquid phase. The N and O_2 contents of soil air are almost the atmospheric air but the concentration of CO_2 is much higher (8 – 10 times more) which may be toxic to plant roots. This phase supplies O_2 and thereby prevents CO_2 toxicity. The 3 phases of the soil system have definite roles to play. The solid phase provides mechanical support for and nutrients to the plants. The liquid phase supplies water and along with it dissolved nutrients to plant roots. The gaseous phase satisfies the acration (O_2) need of plants.

2

Physical Properties of Soil

Soil acts as a habitat for soil organisms, a regulator of water quality, a recycling system for organic wastes and nutrients, a modifier of the composition of the atmosphere, an important provider of ecosystem services and a medium that supports plant growth. These services depend on the physical properties. The diverse physical properties of the soil such as texture, structure, density, porosity, color, resistivity, etc. and their significance have been extensively discussed in this chapter.

Texture

Soil texture is the relative proportions of sand, silt, or clay in a soil. The soil textural class is a grouping of soils based upon these relative proportions. Soils with the finest texture are called clay soils, while soils with the coarsest texture are called sands. However, a soil that has a relatively even mixture of sand, silt, and clay and exhibits the properties from each separate is called a loam. There are different types of loams, based upon which soil separate is most abundantly present. If the percentages of clay, silt, and sand in a soil are known (primarily through laboratory analysis), you may use the textural triangle to determine the texture class of your soil.

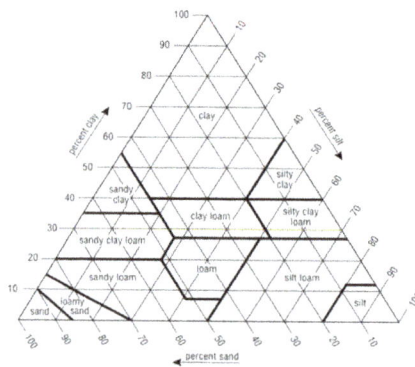

Figure: Textural Triangle. The textural triangle describes the relative proportions of sand, silt and clay in various types of soils.

Using the Triangle

A soil is made up of 50% clay and 20% silt. To work out the soil's texture, follow steps 1-4 and use the diagram.

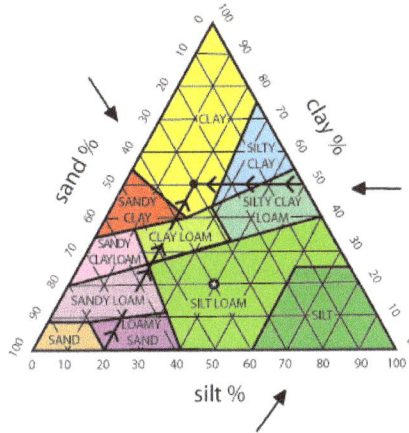

Step One

Find the 50% clay mark on the texture diagram. Start at 0 on the clay line. Move your pencil along until it comes to 50%. Now draw a line from the 50% mark horizontally across and parallel to the base of the triangle.

Step Two

Look at the horizontal line with the silt numbers. Find where 0 is on this line. Move your pen along until it reaches the 20% silt mark. Now draw an upward line parallel to the left hand side of the triangle.

Step Three

The two lines will meet in an area with the name of the soil you have. The soil is clay.

Step Four

You don't need to know the percentage of the third kind of particles in this soil. If you wanted to find out how much sand is in the soil this is what you would do.

100% – 20% silt – 50% clay = 30% sand

Tests to Determine Soil Texture

For fish-pond construction, it is better to have a soil with a high proportion of silt and/or clay which will hold water well. To check quickly on the texture of the soil at different depths, here are two very simple tests you can perform.

Throw-the-ball Test

- Take a handful of moist soil and squeeze it into a ball;

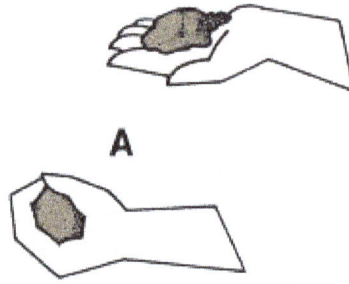

- Throw the ball into the air about 50 cm and then catch it.

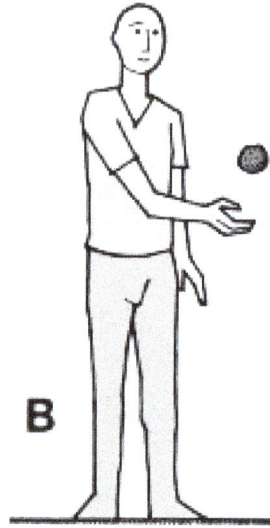

- If the ball falls apart, it is poor soil with too much sand; and

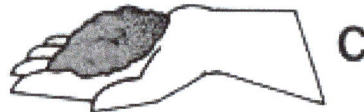

- If the ball sticks together, it is probably good soil with enough clay in it.

Squeeze-the-ball Test

- Take a handful of soil and wet it, so that it begins to stick together without sticking to your hand;

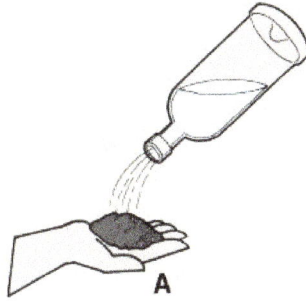

- Squeeze it hard, then open your hand .

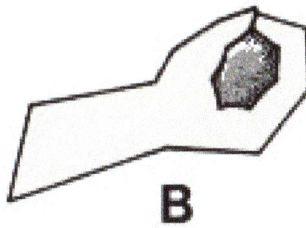

- If the soil retains the shape of your hand, there is probably enough clay in it to build a fish pond;

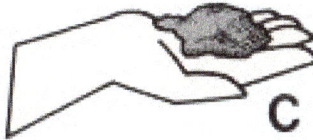

- If the soil does not retain the shape of your hand, there is too much sand in it.

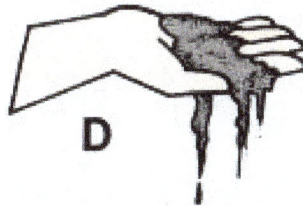

How to Find the Approximate Proportions of Sand, Silt and Clay

This is a simple test which will give you a general idea of the proportions of sand, silt and clay present in the soil.

The Bottle Test

- Put 5 cm of soil in a bottle and fill it with water;

A

- Stir the water and soil well, put the bottle down, and do not touch it for an hour. At the end of an hour, the water will have cleared and you will see that the larger particles have settled;

B

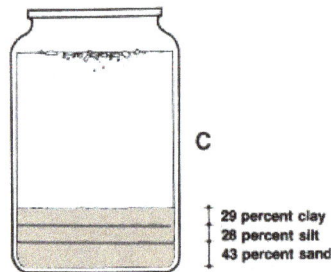

C

29 percent clay
28 percent silt
43 percent sand

- At the bottom is a layer of sand;

- In the middle is a layer of silt;

- On the top is a layer of clay. If the water is still not clear, it is because some of the finest clay is still mixed with the water;

- On the surface of the water there may be bits of organic matter floating;

- Measure the depth of the sand, silt and clay and estimate the approximate proportion of each.

How to Rate Soil Texture From Fine to Coarse

Soil texture may be rated from fine to coarse. A fine texture indicates a high proportion of finer particles such as silt and clay. A coarse texture indicates a high proportion of sand. More precise definitions may be obtained from Table. The simple test below will help you to rate the soil texture from coarse to fine.

The Mud-ball Test

- Take a handful of soil, wet it, and work it to the consistency of dough;

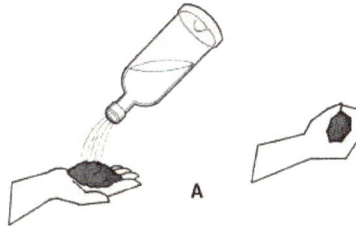

A

- Continue to work it between thumb and forefinger and make a mud ball about 3 cm in diameter;

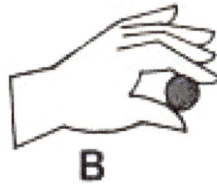

B

- Soil texture can be determined by the way the ball acts when you throw it at a hard surface, such as a wall or a tree .

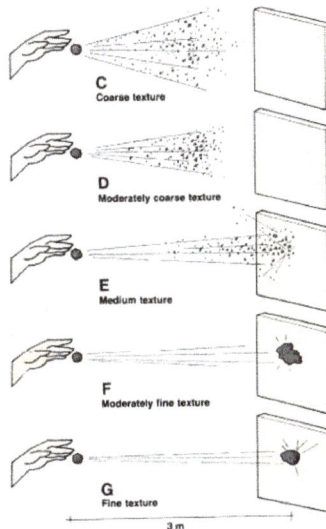

C
Coarse texture

D
Moderately coarse texture

E
Medium texture

F
Moderately fine texture

G
Fine texture

3 m

- If the soil is good only for splatter shots (C) when either wet or dry, it has a coarse texture;

- If there is a "shotgun" pattern (D) when dry and it holds its shape against a medium-range target when wet, it has a moderately coarse texture;

- If the ball shatters on impact (E) when dry and clings together when moist but does not stick to the target, it has a medium texture;

- If the ball holds its shape for long-range shots (F) when wet and sticks to the target but is fairly easy to remove, it has a moderately fine texture;

- If the ball sticks well to the target (G) when wet and becomes a very hard missile when dry, it has a fine texture.

Soil Textural Classes and Field Tests for their Determination

A More Accurate Determination of Soil Texture

Soils may be assigned to textural classes depending on the proportions of sand, silt and clay-size particles. These textural classes are defined in Table 4 and they are represented in Table 6. In the field, there are several ways by which you can find the textural class of the fine-earth portion of a particular soil sample.

The Ball-shaking Test

- Take a handful of soil and wet it;

- Make a ball about 3-5 cm in diameter;

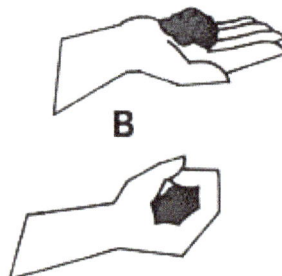

- Place the ball on the palm of your hand: it appears shiny;

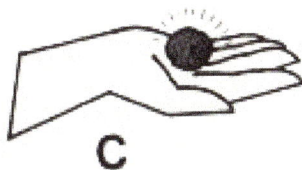

- Shake it from side to side rapidly while watching the surface of the ball .

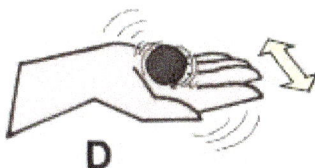

- If the surface of the ball becomes rapidly dull and you can easily break the ball between your fingers, it is sand or loamy sand;

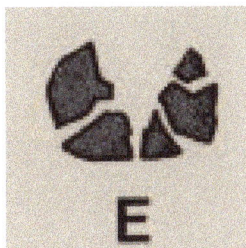

- If the surface of the ball becomes dull more slowly and you feel some resistance when breaking the ball between your fingers, it is silt or clay loam;

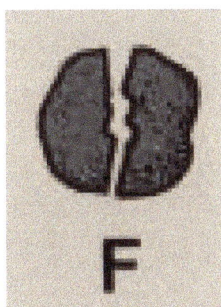

- If the surface of the ball does not change and you feel resistance when breaking the ball, it is clay or silty clay.

The Dry Crushing Test

- Take a small sample of dry soil in your hand;

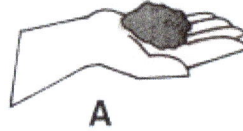

- Crush it between your fingers

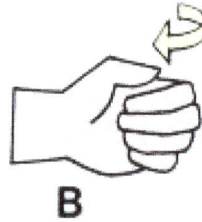

- If there is little resistance and the sample falls into dust, it is fine sand or fine loamy sand or there is very little clay present;

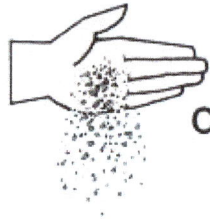

- If there is medium resistance, it is silty clay or sandy clay;

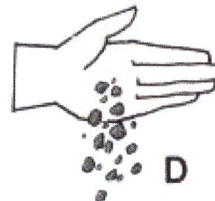

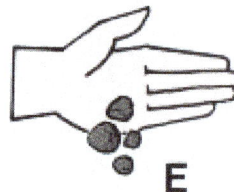

- If there is great resistance, it is clay.

The Manipulative Test

The manipulative test gives you a better idea of the soil texture. This test must be performed exactly in the sequence described below because, to be successful, each step requires progressively more silt and more clay.

- Take a handful of soil and wet it so that it begins to stick together, but without sticking to your hand;

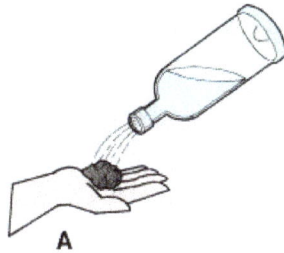

A

- Roll the soil sample into a ball about 3 cm in diameter;

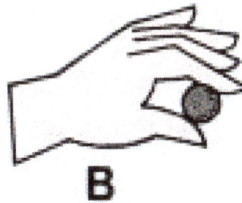

B

- Put the ball down.

C

- If it falls apart, it is sand;
- If it sticks together, go on to the next step.
- Roll the ball into a sausage shape, 6-7 cm long .

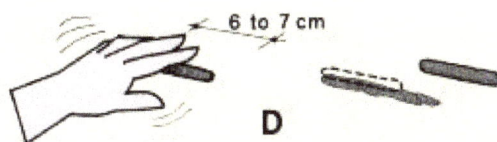

6 to 7 cm

D

- If it does not remain in this form, it is loamy sand;

- If it remains in this shape, go on to the next step.

- Continue to roll the sausage until it reaches 15-16 cm long

- If it does not remain in this shape, it is sandy loam;

- If it remains in this shape, go on to the next step.

- Try to bend the sausage into a half circle .

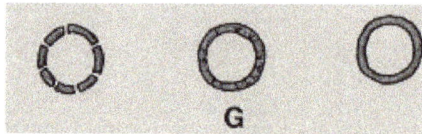

- If you cannot, it is loam;

- If you can, go on to the next step.

- Continue to bend the sausage to form a full circle .

- If you cannot, it is heavy loam;

- If you can, with slight cracks in the sausage, it is light clay;

- If you can, with no cracks in the sausage, it is clay.

The Shaking Test: How to Differentiate Clay From Silt

Both silt and clay soils have a very smooth texture. It is very important to be able to tell the difference between these two soils because they may behave very differently when used as construction material for dams or dikes where the silt may not have enough plasticity. Silty soils when wet may become very unstable, while clay is a very stable construction material. plasticity. Silty soils when wet may become very unstable, while clay is a very stable construction material.

- Take a sample of soil and wet it;

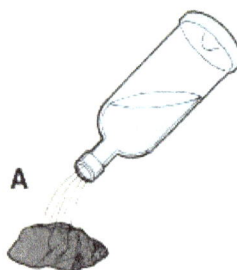

- Form a patty about 8 cm in diameter and about 1.5 cm thick;

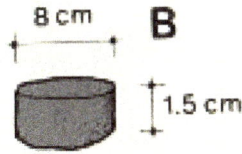

- Place the patty in the palm of your hand: it appears dull;
- Shake the patty from side to side while watching its surface . . .
- If its surface appears shiny, it is silt;
- If its surface appears dull, it is clay.

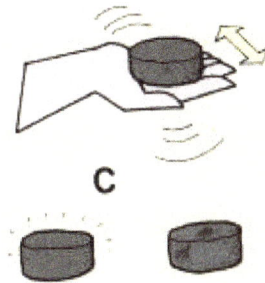

- Confirm this result by bending the patty between your fingers . . .

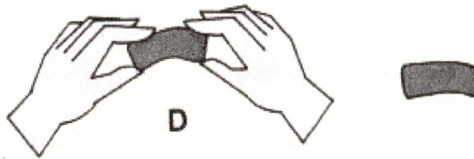

- If its surface becomes dull again, it is silt;
- Put the patty aside and let it dry completely;

- If it is brittle and dust comes off when rubbing it with your fingers, it is silt; and

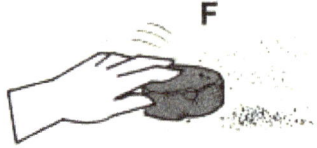

- If it is firm and dust does not come off when rubbing it with your fingers, it is clay.

Note: record the results of the shaking test - rapid, slow, very slow, not at all - according to the speed with which the surface of the patty becomes shiny when you shake it.

Table: USDA textural classes of soils

Common names of soils (General texture)	Sand	Silt	Clay	Textural class
Sandy soils (Coarse texture)	86-100	0-14	0-10	Sand
	70-86	0-30	0-15	Loamy sand
Loamy soils (Moderately coarse texture)	50-70	0-50	0-20	Sandy loam
Loamy soils (Medium texture)	23-52	28-50	7-27	Loam
	20-50	74-88	0-27	Silty loam
	0-20	88-100	0-12	Silt
Loamy soils (Moderately fine texture)	20-45	15-52	27-40	Clay loam
	45-80	0-28	20-35	Sandy clay loam
	0-20	40-73	27-40	Silty clay loam
Clayey soils (Fine texture)	45-65	0-20	35-55	Sandy clay
	0-20	40-60	40-60	Silty clay
	0-45	0-40	40-100	Clay

Structure

Soil structure refers to the arrangement of soil separates into units called soil aggregates. An aggregate possesses solids and pore space. Aggregates are separated by planes of weakness and are dominated by clay particles. Silt and fine sand particles may also be part of an aggregate. The aggregate acts like a larger silt or sand particle depending upon its size.

The arrangement of soil aggregates into different forms gives a soil its structure. The natural processes that aid in forming aggregates are:

1. Wetting and drying,

2. Freezing and thawing,

3. Microbial activity that aids in the decay of organic matter,

4. Activity of roots and soil animals, and

5. Adsorbed cations.

The wetting/drying and freezing/thawing action as well as root or animal activity push particles back and forth to form aggregates. Decaying plant residues and microbial by-products coat soil particles and bind particles into aggregates. Adsorbed cations help form aggregates whenever a cation is bonded to two or more particles.

Ideal Soil Structure

A soil with an ideal structure has properties midway between a sandy soil and a clay soil. It has groups of crumbs about 1mm to 5 mm in size. This occurs naturally in some soils, especially loams that contain the three particles sand, silt and clay.

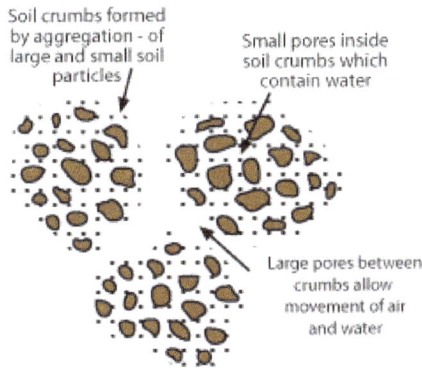

Aggregates are described by their shape, size and stability. Aggregate types are used most frequently when discussing structure.

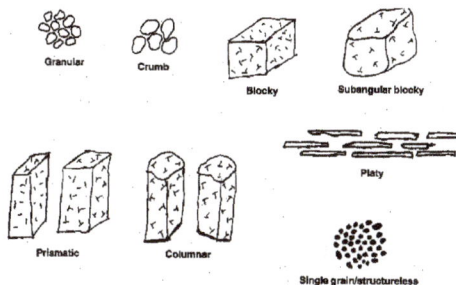

Figure: Soil structural types.

Table: Structure type and description

Type	Description
Granular	Rounded surfaces
Crumb	Rounded surfaces but larger than granular
Subangular blocky	Cube-like with flattened surfaces and rounded corners
Blocky	Cube-like with flattened surfaces and sharp corners
Prismatic	Rectangular with a long vertical dimension and flattened top
Columnar	Rectangular with a long vertical dimension and rounded top
Platy	Rectangular with a long horizontal dimension
Single grain	No aggregation of coarse particles when dry
Structureless	No aggregation of fine particles when dry

Structure is one of the defining characteristics of a soil horizon. A soil exhibits only one structure per soil horizon, but different horizons within a soil may exhibit different structures. All of the soil-forming factors, especially climate, influence the type of structure that develops at each depth. Granular and crumb structure are usually located at the soil surface in the A horizon. The subsoil, predominantly the B horizon, has sub angular blocky, blocky, columnar or prismatic structure. Platy structure can be found in the surface or subsoil while single grain and structure less structure are most often associated with the C horizon. Turn to Soils - Part 1 to identify the structure for different horizons of the Holdrege, Nora, Sharpsburg and Valentine soils.

Aggregates are important in a soil because they influence bulk density, porosity and pore size. Pores within an aggregate are quite small as compared to the pores between aggregates and between single soil particles. This balance of large and small pores provides for good soil aeration, permeability and water-holding capacity.

Tillage, falling raindrops and compaction are primarily responsible for destroying aggregates. As the cutting edge of a tillage implement is pulled through the soil, the shearing action at the point of contact breaks apart aggregates. If tillage is conducted at the same depth for several years, a tillage pan may develop. This is one form of compaction. Particles that were once part of the aggregates may reorient themselves and form platy structures. The amount of aggregate destruction that results from tillage depends on the amount of energy the tillage implement places in the soil. The field cultivator has little down pressure and destroys few aggregates. The disk, however, has both cutting action because of the rotation of the disk and shearing action. Together there is substantial down pressure and destruction.

Aggregates on the soil surface can be broken down by the beating action of raindrops. The single particles that were once part of the aggregate can easily form a crust when the soil dries. The crust looks very similar to the crust formed on a puddle after it rains. It is very difficult for water to infiltrate a crust and for seedlings to push up through a crust. Thus, field operations that lead to aggregate destruction at the soil surface have detrimental secondary effects. The particles also can be eroded if they become detached by rainfall.

Compaction can lead to the breakdown of aggregates in the surface soil and subsoil if the applied force from wheel traffic, animal traffic or human traffic is greater than the force holding an aggregate together. Field observations have shown that compaction can cause granular structure on the soil surface to break down and reform as blocky structure and blocky or sub angular blocky structure in the subsoil to become structure less.

Aggregation is promoted by root growth and the addition of organic material. Roots excrete compounds that are used as food by microorganisms. Also, as roots absorb water and dry the soil, cracks form along planes of weakness. Lastly, when roots decay, root channels serve as conduits for water that facilitate wetting/drying and freezing/thawing.

Organic material may be added in the form of crop residue, animal manure, sludge, and green manure. These additions are usually made to the surface soil and are critical to the development of granular and crumb structure. As organic material is incorporated by tillage, soil animals and microorganisms, it aids in subsoil structure development.

Effect of Soil Structure on Erosion

Soils with a well-developed soil structure are less likely to be eroded by wind and water. Sandy soils that have no structure are easily moved by wind and water. Their surface structure moves easily.

Good sized pore spaces allow water to move through the soil easily, and that water is available to plant roots. This avoids problems of surface erosion.

Density

Density represents weight (mass) per unit volume of a substance.

Density = Mass / Volume

Soil density is expressed in two well accepted concepts as particle density and bulk density. In the metric system, particle density can be expressed in terms of mega grams per cubic meter (Mg/m^3). Thus if 1 m^3 of soil solids weighs 2.6 Mg, the particle density is 2.6 Mg / m^3 (since 1 Mg =1 million grams and 1 m^3 =1 million cubic centimeters) thus particle density can also be expressed as 2.6 g / cm^3.

Types of Density

1. Particle Density:

The weight per unit volume of the solid portion of soil is called particle density. It is also termed as true density. It is expressed in gm./c.c. (C.G.S. system) or lb/cft (F.P.S. system). It depends upon the accumulative densities of the individual inorganic and organic constituents of the soil. Generally in the normal soils the particle density is 2.65 grams per cubic centimetre or mega grams per cubic metre.

The particle density is higher if large amounts of heavy minerals such as magnetite, limonite, hematite and zircon are present. With an increase in organic matter of the soil, the particle density decreases. When particle density is divided by density of water, a relative weight number is obtained is called specific gravity.

Table Particle density of different soil textural classes

Textural classes	Particle density (g/ cm^3)
Coarse sand	2.655
Fine sand	2.659
Silt	2.798
Clay	2.837

Bulk density is an indicator of soil compaction and soil health. It affects infiltration, rooting depth/restrictions, available water capacity, soil porosity, plant nutrient availability, and soil microorganism activity, which influence key soil processes and productivity. It is the weight of dry soil per unit of volume typically expressed in grams/cm^3. Total volume of surface soil is about 50% solids, mostly soil particles (45%), and organic matter (generally < 5%); and about 50% pore space which are filled with air or water. When determining bulk density, the amount of soil moisture must be determined. Available water capacity is the amount of soil moisture available to plants, varies by texture, and is reduced when compaction occurs. Bulk density can be managed, using measures that limit compaction and build soil organic matter.

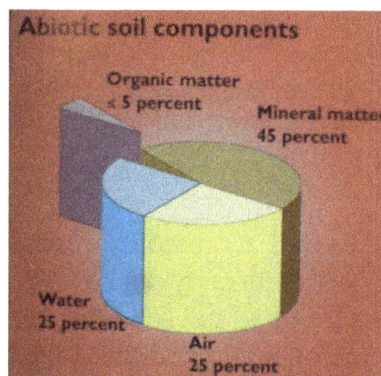

Figure: Four major components of soil volume

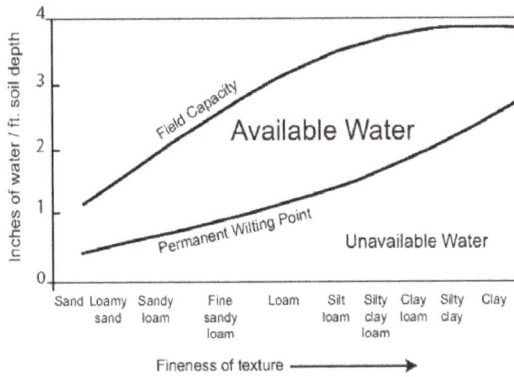

Figure: Relationship between available water and texture

Inherent Factors Affecting Bulk Density and Available Water Capacity

Inherent factors that affect bulk density such as soil texture cannot be changed. Bulk density is dependent on soil organic matter, soil texture, the density of soil mineral (sand, silt, and clay) and their packing arrangement. As a rule of thumb, most rocks have a density of 2.65 g/cm³ so ideally, a silt loam soil has 50% pore space and a bulk density of 1.33 g/cm³. Generally, loose, well-aggregated, porous soils and those rich in organic matter have lower bulk density. Sandy soils have relatively high bulk density since total pore space in sands is less than silt or clay soils.

Bulk density typically increases with soil depth since subsurface layers are more compacted and have less organic matter, less aggregation, and less root penetration compared to surface layers, therefore contain less pore space.

Available water capacity is affected by soil texture, presence and abundance of rock fragments, soil depth and restrictive layers.

Bulk Density Management

Bulk density can be changed by management practices that affect soil cover, organic matter, soil structure, compaction, and porosity. Excessive tillage destroys soil organic matter and weakens the natural stability of soil aggregates making them susceptible to erosion caused by water and wind. When eroded soil particles fill pore space, porosity is reduced and bulk density increases. Tillage and equipment travel results in compacted soil layers with increased bulk density, most notably a "plow pan". Tillage prior to planting temporarily decreases bulk density on the surface but increases at the depth of tillage. Subsequent trips across the field by farm equipment, rainfall events, animals, and other disturbance activities will also compact soil. Long-term solutions to soil compaction centre on decreasing soil disturbance and increasing soil organic matter.

A soil's available water capacity is also affected by organic matter and compaction. Organic matter increases a soil's ability to hold water, both directly and indirectly.

Compaction increases bulk density and reduces total pore volume, consequently reducing available water holding capacity.

The following measures increase organic matter, and reduce compaction, which improve bulk density and porosity:

- Practices that increase organic matter such as continuous no-till, cover crops, solid manure or compost application, diverse rotations with high residue crops and perennial legumes or grass used in rotation;

- Minimize soil disturbance and avoid operating equipment when soils are wet;

- Use designated roads or rows for equipment;

- Reduce the number of trips across a field;

- Subsoil to disrupt existing compacted layers; and

- Use multi-crop systems involving plants with different rooting depths to help break up compacted soil layers.

Figure: Compacted soil in wheel traffic row.

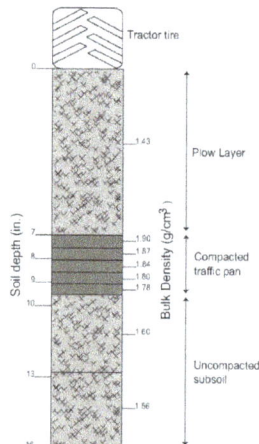

Figure: Compacted plow layer inhibiting root penetration and water movement through soil profile

Water-Filled Pore Space and Porosity

When determining bulk density, water-filled pore space and porosity can also be calculated. When water-filled pore space exceeds 60%, several important soil processes are impacted. Soil respiration and nitrogen cycling (ammonification and nitrification) increase with increasing soil moisture. Lack of aeration also interferes with a soil organism's ability to respire and cycle nitrogen. In dry soils, these processes decline because of lack of soil moisture.

As soil water-filled pore space exceeds 80%, soil respiration declines to a minimum level and denitrification occurs resulting in loss of nitrogen as gas, emission of potent greenhouse gases, yield reduction, and/or increased N fertilizer expense.

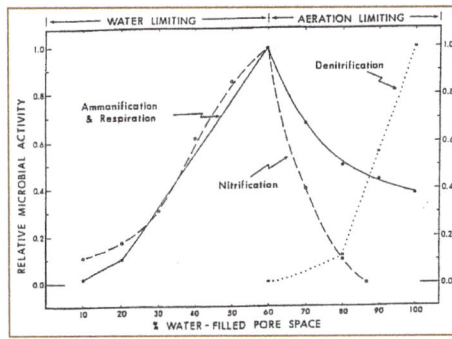

Figure: Relationship of water-filled pore space to soil microbial activity

Soil Bulk Density Problems and Relationship to Soil Function

High bulk density is an indicator of low soil porosity and soil compaction. High bulk density impacts available water capacity, root growth, and movement of air and water through soil. Compaction increases bulk density and reduces crop yields and vegetative cover available to protect soil from erosion. By reducing water infiltration into soil, compaction can lead to increase runoff and erosion from sloping land or saturated soils in flatter areas. Bulk densities above thresholds in Table impair root growth.

Laboratory analyses should use bulk density to adjust volume of soil to determine organic matter and nutrient content of soil. For example, a soil with a bulk density of 1.3 versus a soil with a bulk density of 1.0 would result in a 30% error in organic matter and nutrient content if the samples are not adjusted for bulk density.

Table: General relationship of soil bulk density to root growth based on soil texture.

Soil Texture	Ideal bulk densities for plant growth (grams/cm^3)	Bulk densities that affect root growth (grams/cm^3)	Bulk densities that restrict root growth (grams/cm^3)
Sands, loamy sands	< 1.60	1.69	> 1.80
Sandy loams, loams	< 1.40	1.63	> 1.80

Sandy clay loams, clay loams	< 1.40	1.60	> 1.75
Silts, silt loams	< 1.40	1.60	> 1.75
Silt loams, silty clay loams	< 1.40	1.55	> 1.65
Sandy clays, silty clays, clay loams	< 1.10	1.49	> 1.58
Clays (> 45% clay)	< 1.10	1.39	> 1.47

Measuring Bulk Density and Soil Moisture

Materials Needed to Measure Bulk Density

- 3-inch diameter aluminum ring
- Wood block or plastic insertion cap
- Rubber mallet or weight
- Folding trowel
- Flat-bladed knife
- Sealable bags and marker pen
- Scale (1 g precision)
- 1/8 cup (29.5 mL) measuring scoop
- Ceramic coffee cup or paper plate
- 18-inch metal rod, probe or spade (to check for compaction zone)
- Access to a microwave oven

Considerations – Bulk density can be measured at the soil surface and/or compacted tillage zone. Bulk density samples should be taken in same location as infiltration and respiration tests. It may be possible to use the infiltration test sample. For sticky clay soils a little penetrating oil applied to the ring makes it easier to remove soil.

Step by Step Procedure

Figure: Drive ring to 3-inch depth.

1. Carefully clear all residue then drive ring to a depth of 3 inches (2 inches from top) with small mallet or weight and block of wood or plastic cap (same process as used for infiltration test).

Figure: Ring driven to 3-inch depth.

2. Remove ring by cutting around the outside edge with a small 4-inch serrated butter knife and using the small folding trowel underneath of it and carefully lift the ring out preventing loss of soil by holding trowel under it.

3. Remove excess soil from the bottom of cylinder with serrated butter knife as shown in Figure.

Figure: Remove excess soil from bottom of ring.

4. Place sample in plastic sealable bag and label it.

5. Weigh sample in bag and record its weight in Table.

6. Weigh an identical clean, empty plastic bag and record its weight in Table.

7. Weigh empty cup or paper plate to be used in step 8 and record its weight on Table 2.

8. Either extract a subsample shown in steps 8-10, or dry and weigh entire sample to determine water content and dry soil weight:

a. Mix sample thoroughly in the bag by kneading it with your fingers.

b. Take a 1/8-cup level scoop of loose soil (not packed down) from plastic bag and place it in the cup weighed in step 7 (use more than one scoop to increase accuracy).

9. Weigh moist subsample in cup before drying and record in table.

10. Place cup containing subsample in a micro wave and dry for two or more four-minute cycles at medium power.

11. To determine if soil is dry, weigh subsample in cup/plate after each 4-minute cycle. When the weight no longer changes after a drying cycle, it is dry, and record its weight on Table.

Porosity

Soil porosity, or soil pore space, are the small voids between particles of soil. In heathy soil, these pores are large and plentiful enough to retain the water, oxygen and nutrients that plants need to absorb through their roots. Soil porosity usually falls into one of three categories: micro-pores, macro-pores or bio-pores.

Types of Pores in Soil

In general there are broadly two types of pores in soils:

1. Macro-pores and

2. Micro or capillary pores.

(i) Macro Pores:

Large-sized pores are referred to as macro-pores which allow air and water movement easily. Sands and sandy soils have a large number of macro-pores. It is found in between the granules.

(ii) Micro or Capillary Pores:

Smaller sized pores are generally referred to as a micro or capillary pores in which movement of air and water is restricted to some extent. Clays and clayey soils have a greater number of micro or capillary pores. It has got more important in the plant growth relationship. It is found within the granules.

Besides soil pores have been divided into following four categories based on the size groupings of soil separates:

1. Coarse Pores:

Greater than 0.2 mm or 200 microns (0.008 inch) average diameter.1 micron = 1 millionth of a metre.

2. Size of Medium Sands:

a. Medium Pores:

0.2-0.02 mm or 200-20 microns (0.008-0.0008 inch) average diameter. Size of coarse silt particles.

b. Fine Pores:

0.02-0.002 mm, which is 20-2 microns (0.0008 inch) average diameter? Size of fine silt particles.

c. Very Fine Pores:

Less than 2 microns (0.00008 inch) average diameter (Large clay particles are 2 microns in average diameter). Size of large clay particles. Porosity refers to the percentage of soil volume occupied by pore spaces. Size of individual pores, rather than total pore space in a soil, is more significant in its plant growth relationship.

For optimum growth of the plant, the existence of approximately equal amount of macro and micro-pores which influence aeration, permeability, drainage and water retention favorably. Porosity of a soil can be easily changed.

Factors Affecting Porosity of Soil

Wide difference in the total pore space of various soils occurs depending upon the following several factors:

(i) Soil Structure:

A soil having granular and crumb structure contains more pore spaces than that of prismatic and platy soil structure. So well aggregated soil structure has greater pore space as compared to structure less or single grain soil.

(ii) Soil Texture:

In sandy soils the total pore space is small whereas in fine textured clay and clayey loam soils total pore space is high and there is a possibility of more granulation in clay soils.

(iii) Arrangement of Soil Particles:

When the sphere like particles is arrangement in columnar form (i.e. one after another on the surface forming column like shape) it gives the most open packing system resulting very low amount of pore spaces. When such particles are arranged in the pyramidal form it gives the most close packing system resulting high amount of pore spaces.

(iv) Organic Matter:

Soil containing high organic matter possesses high porosity because of well aggregate formation.

(v) Macro-Organisms:

Macro-organisms like earthworm, rodents, and insects etc. increase macro-pores in the soil.

(vi) Depth of Soil:

With the increase in depth of soil, the porosity will decrease because of compactness in the sub-soil.

(vii) Cropping:

Intensive crop cultivation tends to lower the porosity of soil as compared to fallow soils. The decrease in porosity may be due to reduction in organic matter content.

(viii) Puddling:

Due to puddling under sufficient soil moisture, the soil surface layer is made dense and compact. Eventually, the porosity of this surface soil is reduced by the infiltration of muddy surface materials.

Things that Makes Soil Porous

While the small micro-pores of clay soil can retain water and nutrients longer than sandy soil, the pores themselves are often too small for the plant roots to be able to properly absorb them. Oxygen, which is another important element needed in soil pores for proper plant growth, may also have a hard time permeating clay soils. In addition, compacted soils have decreased pore space to hold the necessary water, oxygen and nutrients needed for developing plants.

This makes knowing how to get porous soil in the garden important if you want healthier plant growth. So how can we create healthy porous soil if we find ourselves with clay-like or compacted soil? Usually, this is as simple as thoroughly mixing in organic material such as peat moss or garden gypsum to increase soil porosity.

When mixed into clay soil, for instance, garden gypsum or other loosening organic materials can open up the pore space between soil particles, unlocking the water and nutrients that had become trapped in the small micro-pores and allowing oxygen to penetrate the soil.

Porosity Model

The pore spaces of a set of sandstone samples are fractals and self-similar over a definite range. Their models correctly predicted porosity and the expression used was

$$\phi = A\left(L_1 / L_2\right)^{3-D},$$

where φ is the porosity of sandstone, A is a constant which is equal to 1 over some range, L_1 is the measurement size which is equal to the minimum size of a crystal nucleus corresponding to the total porosity, L_2 is the length of rocks or soils under consideration and D is the volume fractal dimension of pores. This equation has been viewed sceptically by experts on fractals because of its lack of provenance. The porosity expressions established by the workers in two- and three-dimensional spaces are respectively

$$\phi = \left(L / L_2\right)^{2-D},$$

and

$$\phi = \left(L / L_2\right)^{3-D},$$

Where, L is the measurement size which can be considered as the particle-size. Virtually, the direct research objects of above equations are the grains of rocks or soils and the background objects are the pores above equations show that the porosity decreases as the grains get smaller. The grains of particle-size smaller than the measurement scale being used are considered as pores, and thus the porosity predictions from above equations are always larger than the actual porosity. Clearly, the larger the measurement scale is, the larger the errors in above equations are. Only when the measurement scale is equal to the minimum particle-size, will the total porosity predictions from above equations match reality. Overall, above equations cannot be used to accurately describe the pore-size distribution of rocks and soils.

Based on a theoretical sequence of soil aggregates, an equation describing the porosity of soil aggregates was derived by Rieu and Sposito and the expression is

$$\phi = (d_{min} / d_{max})^{3-D}$$

where d_{min} and d_{max} are the smallest and the largest size of soil aggregates, respectively, and the definition of D is the fractal dimension of a porous medium. Obviously, when the fractal dimension D increases and approaches 3, the porosity φ in equation ($\phi = (d_{min} / d_{max})^{3-D}$) approaches 0. Yu showed that equation ($\phi = (d_{min} / d_{max})^{3-D}$) is inconsistent with the physical situation. He concluded that when the fractal dimension approaches 3, the space is occupied by pores and the porosity φ would be equal to 1. In fact, Yu considers that the fractal dimension D in equation ($\phi = (d_{min} / d_{max})^{3-D}$) refers to the volume fractal dimension of pores, but that the fractal dimension is actually the volume fractal dimension of grains. When the volume fractal dimension of grains increases and approaches 3, the space is occupied by grains and the porosity φ would be equal to 0, then equation ($\phi = (d_{min}/d_{max})^{3-D}$) represents the physical situation. The definition of the volume (area) fractal dimension of grains (pores) in this study is similar to. However, there is still no direct evidence to support equation ($\phi = (d_{min}/d_{max})^{3-D}$), so equation ($\phi = (d_{min}/d_{max})^{3-D}$) cannot be confirmed to be accurate. In addition, equation ($\phi = (d_{min}/d_{max})^{3-D}$) gives the relationship between porosity and aggregate size, but not one for porosity and pore-size, thus it is not an ideal model that can be used to accurately describe the pore distribution of rocks and soils. Furthermore, Rieu & Sposito considered that D in equations ($\phi = (d_{min}/d_{max})^{3-D}$) can be calculated through the logarithmic relationship between the bulk density of soil aggregates and aggregate size, or the number of soil aggregates and aggregate size. In fact, the fractal dimension calculated by this means does not accurately reflect the pore-size distribution of rocks and soils, because this fractal dimension is not based on experimental data of pore-size distribution and the total area (volume) of rocks and soils.

Consistency

Soil consistency is the strength with which soil materials are held together or the resistance of soils to deformation and rupture. Soil consistency is measured for wet, moist and dry soil samples. For wet soils, it is expressed as both stickiness and plasticity, as defined below. Soil consistency may be estimated in the field using simple tests or may be measured more accurately in the laboratory.

Wet Soils

In wet soils the consistency is denoted by terms stickiness and plasticity. Stickiness is grouped into four categories namely i) non-sticky, ii) slightly sticky, iii) sticky and iv) very sticky. Plasticity of a soil is its capacity to be moulded (to change its shape depending on stress) and to retain the shape even when the stress is removed. Soils containing more than 15% clay exhibit plasticity – pliability and the capacity of being molded. There are four degrees in plasticity namely i) non-plastic, ii) slightly plastic, iii) plastic and iv) very plastic.

Determination of Wet-soil Consistency

Testing is done when the soil is saturated with water, as, for example, immediately after a good rainfall. First, determine stickiness, that is, the ability of soil materials to adhere to other objects. Then, determine plasticity, that is, the ability of soil materials to change shape, but not volume, continuously under the influence of a constant pressure and to retain the impressed shape when the pressure is removed.

Field Test for Stickiness of Wet Soil

Press a small amount of wet soil between your thumb and forefinger to see if it will stick to your fingers. Then slowly open your fingers. Rate the stickiness as follows:

0. Non-sticky, if no soil or practically no soil sticks to your fingers;

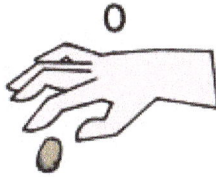

1. Slightly sticky, if the soil begins to stick to your fingers but comes off one or the other cleanly and does not stretch when the fingers are opened;

2. Sticky, if the soil sticks to both the thumb and forefinger and tends to stretch a little and pull apart rather than pulling free from your fingers;

3. Very sticky, if the soil sticks firmly to both thumb and forefinger and stretches when the fingers are opened.

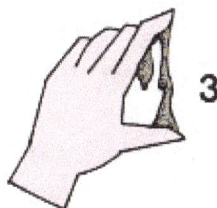

Field Test for Plasticity of Wet Soil

Roll a small amount of wet soil between the palms of your hands until it forms a long, round strip like a wire about 3 mm thick. Rate the plasticity as follows:

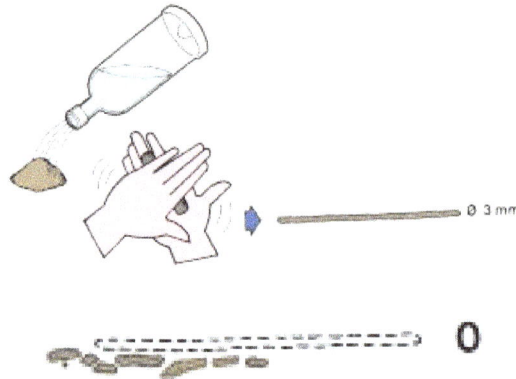

0. Non-plastic, if no wire can be formed;

1. Slightly plastic, if a wire can be formed but can easily be broken and returned to its former state;

2. Plastic, if a wire can be formed but, when it is broken and returned to its former state, it cannot be formed again;

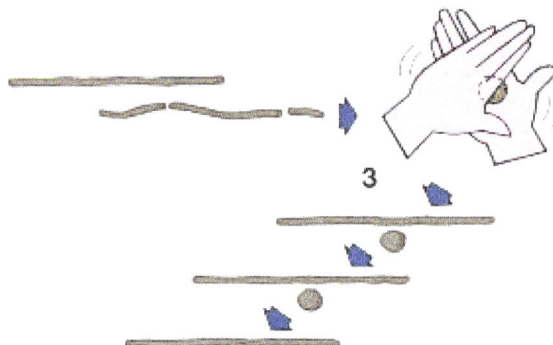

3. Very plastic, if a wire can be formed which cannot be broken easily and, when it is broken, it can be rolled between your hands and be reformed several times.

Moist Soil

Moist soil with least coherence adheres very strongly and resists crushing between the thumb and forefinger. The different categories are i. Loose-non coherent, ii. Very friable - coherent, but very easily crushed, iii. Friable - easily crushed, iv. Firm - crushable with moderate pressure, v. Very firm - crushable only under strong pressure and vi. Extremely firm - completely resistant to crushing. (type and amount of clay and humus influence this consistency).

Determination of Moist-soil Consistency

Field Test for Moist-soil Consistency

Testing is done when the soil is moist but not wet, as, for example, 24 hours after a good rainfall.

Try to crush a small amount of moist soil by pressing it between your thumb and forefinger or by squeezing it in the palm of your hand. Rate moist soil consistency as follows:

0. Loose, if the soil is non-coherent (single-grain structure);

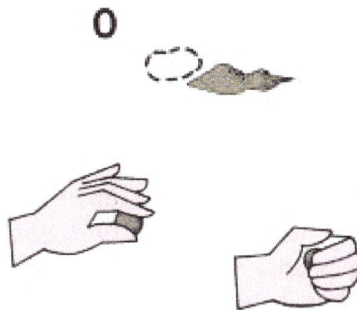

1. Very friable, if the soil crushes easily under very gentle pressure but will stick together if pressed again;

2. Friable, if the soil crushes easily under gentle to moderate pressure;

3. Firm, if the soil crushes under moderate pressure but resistance is noticeable;

4. Very firm, if the soil crushes under strong pressure, but this is difficult to do between the thumb and forefinger;

5. Extremely firm, if the soil crushes only under very strong pressure, cannot be crushed between the thumb and forefinger, but must be broken apart bit by bit.

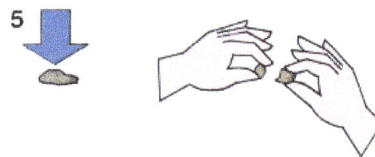

Dry Soil

In dry soil, the degree of resistance is related to the attraction of particles for each other. The different categories are, Loose - non coherent, Soft - breaks with slight pressure and becomes powder, Slightly hard - break under moderate pressure, Hard - breaks with difficulty with pressure, Very hard - very resistant to pressure, Extremely hard - extreme resistance and cannot be broken.

Determination of Dry-soil Consistency

Field Test for Dry-soil Consistency

Testing is done when the soil has been air-dried.

Try to break a small amount of dry soil by pressing it between your thumb and forefinger or by squeezing it in the palm of your hand. Rate dry soil consistency as follows:

0. Loose, if the soil is non-coherent (single-grain structure):

1. Soft, if the soil is very weakly coherent and friable. breaking to powder or individual grains under very slight pressure;

2. Slightly hard, if the soil resists light pressure, but can be broken easily between thumb and forefinger;

3. Hard, if the soil resists moderate pressure, can barely be broken between the thumb and forefinger, but can be broken in the hands without difficulty;

4. Very hard, if the soil resists great pressure, cannot be broken between the thumb and forefinger but can be broken in the hands with difficulty;

5. Extremely hard, if the soil resists extreme pressure and cannot be broken in the hands.

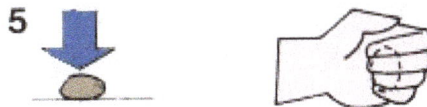

Types of Consistency Limits

In 1911, Atterberg, a Swedish agricultural engineer, stated that a fine-grained soil can exist in four states, namely, liquid, plastic, semi-solid, or solid state, depending on its water content. The water contents at which the soil changes from one state to the other are known as consistency limits or Atterberg limits.

From geotechnical engineering view point, the following three consistency limits are significant:

1. Liquid limit (LL).

2. Plastic limit (PL).

3. Shrinkage limit (SL).

The water content at which the soil changes from the liquid state to the plastic state is known as liquid limit and the water content at which the soil changes from the plastic state to the semi-solid state is known as plastic limit. Similarly, the water content at which the soil changes from the semi-solid state to the solid state is known as shrinkage limit.

Thus, when the water content of a given soil is more than its liquid limit, the soil will be in liquid state and has negligible shear strength. When the water content is between the liquid limit and the plastic limit, the soil exhibits plasticity, that is, the soil can be molded to any shape without rupture. Plastic limit is the minimum water content at which the soil exhibits plasticity.

The shrinkage limit is the water content below which the soil does not undergo any shrinkage and is in solid state. The three consistency limits are shown in figure.

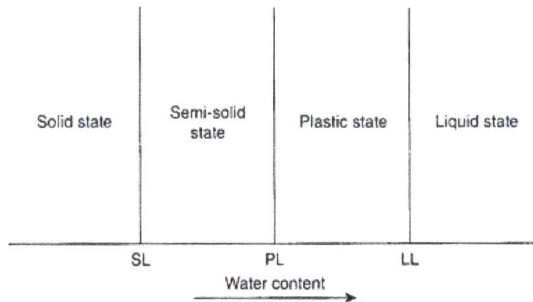

Figure: Consistency of soils as a function of water content.

The liquid limit and plastic limit form the basis for classification of fine-grained soils and for classification of coarse-grained soils with fine fraction. They are also used directly in specifications for controlling compaction used for the construction of embankments and Earth dams. The consistency limits have also been related to various other properties of soils.

1. Liquid Limit

Liquid limit is the water content at which a soil changes from the liquid state to the plastic state. It is the minimum water content at which the soil is still in the liquid state but possesses small shear strength against flow.

The liquid limit is not the same for all soils. The magnitude of liquid limit for a given soil depends on the type and proportion of clay minerals present in the soil. The liquid limit is high for soils containing montmorillonite clay mineral, medium for soils containing illite, and minimum for soils containing kaolinite clay mineral. In general, the higher the clay content present in a given soil, the higher is the liquid limit.

The shear strength of all soils is found to be the same, equal to about 27 g/cm^2 or 0. 027 kgf/cm^2, when their water content is equal to liquid limit. The liquid limit is an important index property of fine-grained soils and helps predict its swelling and compressibility behavior under loads. In general, the higher the liquid limit, the higher will be the swelling of the soil on wetting.

The compressibility, which is the decrease in volume under loads and which determines the settlement of structures, also increases with the increase in the liquid limit of the soil. Skempton gave an equation for compression index in the terms of liquid limit.

The liquid limit of soils can be determined by any one of the following two methods:

1. Casagrande's mechanical method.

2. Uppal's cone penetration method.

2. Plastic Limit

Plastic limit is the water content at which a soil changes from the plastic state to the semi-solid state. It is the minimum water content at which the soil remains in plastic state and can be molded to any shape without rupture.

Principle

Experimentally the plastic limit is defined as the water content at which a soil begins to crumble (forms cracks) when rolled into a thread of 3-mm diameter.

Apparatus

The apparatus consists of a flat square glass plate of minimum 45-cm size and 10-mm thickness, a rod of 3-mm diameter, oven, and containers for water content determination.

Procedure

Following is the procedure for the determination of plastic limit of the soil:

1. About 60 g of air-dried soil passing through 425 pm IS sieve is taken and mixed with sufficient water such that its water content is more than the estimated plastic limit and such that the soil becomes plastic enough to be easily molded with fingers.

2. About 20 g of the thoroughly mixed soil is taken. A ball is made with about 8 g of this soil and rolled on the glass plate with fingers with just sufficient pressure to roll the mass into a thread of uniform diameter, throughout its length.

3. The rate of rolling with fingers shall be at the rate of 80-90 strokes per minute, counting a stroke as one complete forward and backward motion of the fingers.

4. When the diameter of the soil thread reaches 3 mm, the soil thread is worked back to form a ball.

5. The procedure of rolling into a thread of uniform diameter of 3 mm and kneading back into a ball is repeated until cracks appear on the surface of the soil thread, which begins to crumble. When this condition is reached, the water content of the pieces of soil thread is determined.

6. The test is repeated taking another portion of the soil paste; a total of three trials are made and the corresponding water contents are determined. The average water content out of three trials to the nearest whole number is reported as the plastic limit.

3. Shrinkage Limit

Shrinkage limit is the water content at which the soil changes from the semi-solid state to the solid state. For fine-grained soils, it was observed that a decrease in the water content causes a corresponding decrease in the volume of the soil, when the soil is in plastic or semi-solid state. At some water content, a further reduction of the water content does not cause any decrease in the volume of the soil, as shown in figure below. The decrease in the volume of the soil with the decrease in the water content due to drying/evaporation is called as shrinkage.

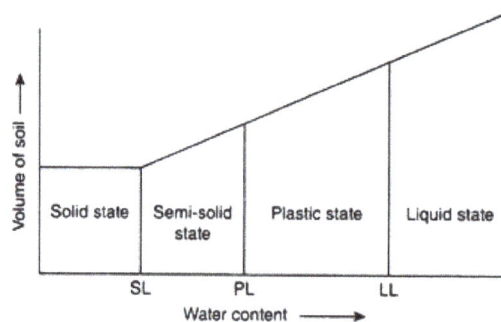

Figure:Shrinkage of a soil as a function of water content.

Uses of Consistency Limits:

Consistency limits are very significant in the study of clays and other fine-grained soils. Important deductions can be made based on the relative values of consistency limits and the index properties as follows:

1. The liquid limit and plasticity index are extremely useful for the classification of soils. The plasticity index and the plasticity chart are used to classify the coarse-grained soils having some fine fraction. The liquid limit and the plasticity chart are used to classify fine-grained soils.

2. The liquid limit is a measure of the compressibility of the soil, that is, the decrease in volume of saturated soils under loads from the structure. Skempton related compression index of soils to the liquid limit.

3. The liquid limit of a clay decreases at a faster rate compared to plastic limit with the increase in silt content in the clay. The plasticity index, therefore, decreases with the increasing silt content in a clayey soil.

4. The liquid limit and plasticity index are an indication of the type and amount of clay present in a soil. The higher the liquid limit and plasticity index, the more severe will be the anticipated problems due to compressibility, swelling, and shrinkage to the foundations and to the structure located in such soils.

5. The compressibility of the soil having higher liquid limit is more, and its shear strength is also less compared to a soil with lower liquid limit, although their plasticity index is the same.

6. Soils with higher plasticity index have a higher dry strength and lower permeability compared to soils with lower plasticity index, although their liquid limit is the same.

7. The plasticity index of the soil increases with the increase in organic content in the soil.

8. Consistency index is a measure of the shear strength of the soils. The higher the consistency index, the higher will be the unconfined compressive strength of the soil.

9. The toughness index is a measure of the shear strength of the soil at its plastic limit and for soils with same plasticity index, the toughness is inversely proportional to flow index.

10. Soils with a high flow index lose their shear strength rapidly with the increase in water content. Such soils cannot sustain heavy loads at high water content.

11. Soils with higher flow index (steeper flow curve) will have lower shear strength if the plasticity index is the same.

12. Earth work can be carried out easily with the least effort when the water content of the soil is at plastic limit.

13. The shrinkage limit is useful to estimate the expected settlements of structures due to drying of soils with change of seasons. The lower the shrinkage limit, the

higher will be the possible settlements of structures. Expansive soils also have low values of shrinkage limit.

14. The plasticity index is a useful measure to verify the suitability of clay for potteries, for the construction of the clay core in an earth dam, and for the construction of a clay liner to contain a polluted material. In all these cases, a high plasticity index is desirable for better workability and low permeability.

Temperature

Soil temperature is the factor that drives germination, blooming, composting, and a variety of other processes. Soil temperature is simply the measurement of the warmth in the soil. Ideal soil temperatures for planting most plants are 65 to 75° F. (18 to 24° C.). Night-time and daytime soil temperatures are both important.

When are soil temperatures taken? Soil temperatures are measured once soils are workable. In zones with higher numbers, the soil temperature will warm up quickly and earlier in the season. In zones that are lower, the soil temperature may take months to warm up as winter chill wears off.

Soil temperature gauges or thermometers are the common way to take the reading. There are special soil temperature gauges used by farmers and soil sample companies, but you can just use a soil thermometer.

In a perfect world, you would check night-time temperatures to ensure they are not so cold your plant's health will be impacted. Instead, check in the early morning for a good average. The night's coolness is still mostly in the soil at this time.

Soil readings for seeds are done in 1 to 2 inches of soil. Sample at least 4 to 6 inches deep for transplants. Insert the thermometer to the hilt, or maximum depth, and hold it for a minute. Do this for three consecutive days. Determining soil temperatures for a compost bin is also done in the morning. The bin should maintain at least 60° F. (16° C.) bacteria and organisms to do their work.

Ideal Soil Temperatures for Planting

The perfect temperature for planting varies dependent upon the variety of vegetable or fruit. Planting before it is time can reduce fruit set, stunt plant growth and prevent or reduce seed germination.

Plants such as tomatoes, cucumbers and snap peas benefit from soils at least 60° F (16 °C.).

Sweet corn, lima beans and some greens need 65 degrees F. (18° C.)

Warmer temperatures into the 70s (20s° C.) are required for watermelon, peppers, squash, and at the higher end, okra, cantaloupe and sweet potatoes.

If you are in doubt, check your seed packet for ideal soil temperatures for planting

Realistic Soil Temperatures

Somewhere between the minimum soil temperature for plant growth and the optimum temperature is the realistic soil temperature. For instance, plants with higher temperature needs, such as okra, have an optimum temperature of 90° F. (32° C.). However, healthy growth can be achieved when they are transplanted into soils of 75 F. (24° C).

This happy medium is suitable for beginning plant growth with the assumption that optimum temperatures will occur as the season progresses. Plants set out in cool zones will benefit from late transplanting and raised beds, where soil temperatures warm up more quickly than ground level planting.

Factors Influencing Soil Temperature

Zhang divided these factors into two; the amount of heat made available to the soil surface and the amount of heat dissipated from soil surface down the profile.

A. The factors that affect the amount of heat supplied at the soil surface include; soil colour, mulching, solar radiation, slope of land surface, vegetative cover, organic matter content and evaporation.

1. Soil colour: Dark coloured soils absorb more radiant heat than light coloured soils. Thus dark coloured soils have a higher soil temperature than light coloured soils.

2. Soil mulch: Observed that mulch materials inhibit evaporation and increase soil moisture. Consequently, these materials reduce the temperature on the soil surface. Therefore, mulching the surface of the soil serves to insulate heat thereby reducing soil temperature. Generally less heat will flow into a mulched soil compared to bare soils.

3. Slope of the land surface: Solar radiation that reaches the land surface at an angle is scattered over a wider area than the same amount of solar radiation reaching the surface of the land at right angles. Therefore the amount of radiation per unit area of the land surface decreases as the slope of the land. Thus soil temperature decreases as the slope of land increases.

4. Vegetative cover: A bare soil quickly absorbs heat, becomes hot during the hot season and becomes cold during the cold season. Vegetation acts as a thermal insulator and significantly affects the soil temperature . It does not allow the soil to become either too hot during the dry season or too cold during the rainy.

5. Organic matter content: Organic matter increases the water holding capacity of the soil. It also contributes to the dark colour of the soil. These two soil properties increase its absorption of heat, thereby increasing the soil.

6. Evaporation: The evaporation of water from the soil requires a large amount of energy. Soil water utilises the energy from solar radiation to evaporate thereby rendering it unavailable for heating up of the soil. Thus the greater the rate of evaporation, the more a soil is cooled and its temperature decreases.

7. Solar radiation: This is the amount of heat from the sun that reaches the earth. The amount of radiation from the sun that a soil receives and absorbs affects the variability of soil temperature. As the solar radiation reaching the soil surface increases the soil temperature also increases.

B. Some of the factors that affect the amount of heat dissipated from the soil down the profile include; moisture content and bulk density.

1. Soil moisture content: Moisture influences soil heat dissipation down the profile. The flow of heat is higher in a wet soil than in a dry soil where the pores are filled with air. The rate of heat dissipation increases with moisture.

2. Bulk density: High bulk density increases the soil surface by increases the amount of heat dissipated through the soil surface by increasing the rate at which heat energy passes through a unit cross-sectional area of the soil.

Effects of Soil Temperature on Some Soil Properties and Plant Growth

Effects on Soil Biological Properties

- Bioactivity: Soil temperature range of 10°C-28°C influence soil respiration by increasing the activity of extracellular enzymes that degrade polymeric organic matter in soils, increase microbial retake of soluble substrates, and increase microbial respiration rates. Increase in soil temperature increases the soil nitrogen mineralization rates through the increase in microbial activity and increase in the decomposition of organic matter in the soil. Soil temperature below freezing point decreases mineralization by inhibiting microbial activity and decreasing diffusion of soluble substrates in the soil.

- Soil micro-organisms: Most soil micro-organisms require temperatures between 10°C -35.6°C for their activities. Soil microbial activities decrease with low soil temperatures and at freezing point.

- Soil macro-organisms: At a soil temperature range of 10°C24°C, soil macro-organisms have increased rate of metabolism requiring them to either feed more or burn their own fat stores. At extreme high temperatures of 58°C, soil macro-organisms die because of the unfavourable temperature of the soil. Soil macro-organisms do not survive in temperatures below freezing point.

Organic matter decomposition: At a temperature below 0°C the accumulation of soil matter increases due to the slow rate of decomposition. Organic matter decomposes slowly at lower temperatures as a result of decrease in microbial activities and biochemical processes. Soil temperature between 2°C-38°C increases the organic matter decomposition by increasing the movement of soluble substrates in the soil and stimulating microbial activities.

Effects on Chemical Properties of the Soil

Cation Exchange Capacity (CEC)

Increase in soil temperature decreases organic matter through combustion. This decrease in organic matter and reduction in clay size clay fraction as a result of high temperature leads to a decrease in the cation exchange capacity of the soil. Increase in soil temperature leads to a decrease in the cation exchange capacity of the soil.

Available Phosphorus

Water-soluble phosphorus increased with soil temperature from $5° c - 25° c$ due to the increase in the movement of phosphorus in the soil controlled by diffusion. Soils with low temperature have low availability of phosphorus because the release of phosphorus from organic material is hindered by low temperature.

Soil pH

At a soil temperature ranges of 25°C- 39°C the soil pH increases as a result of organic acid denaturation which increases at high temperature.

Effects on Soil Physical Properties

Soil Structure

Increase in soil temperature causes temperature induced dehydration of 2:1 clay minerals in the soil leading to strong interactions among the clay particles which in turn yield less clay and more silt-sized particles in the soil. High soil temperature also leads to heat-induced cracks in the sand-sized particles that eventually lead to breakdown and consequently a reduced amount of sand-sized particles in the soil. Increase in soil temperature lowers the clay sand contents and increases the silt content.

Aggregate Stability

At soil temperature above 30° c, the aggregate stability of the soil increases. This is as a result of thermal transformation of iron and aluminium oxides, causing them to act as cementing agents for clay particles that then form strong silt sized particles in the soil.

Soil Moisture Content

Reductions in soil moisture occur when increased soil temperatures decrease water viscosity, thus allowing more water to percolate through the soil profile. In addition, reduced shade combined with increased soil temperatures also results in higher evaporation rates which in turn restrict the movement of water into the soil profile.

Soil Aeration

Temperature influences the carbon dioxide content in the soil air. High temperature encourages micro-organism activity which results in higher production of carbon dioxide in the soil.

Effects of Soil Temperature on Plant Growth

Soil temperature has a great effect on plant growth by influencing water and nutrient uptake, root and shoots growth.

Water Uptake

Water uptake decreases with low temperature. This is due to the increased viscosity and decreased absorption rate of water at low temperature. Decreased water uptake reduces the rate of photosynthesis.

Nutrient Uptake

The metabolic activities of micro-organisms play an important role in the cycling of nutrients in the soil and ensuring the nutrients are in a form available to plants. Therefore increased metabolic activities of micro-organisms as a result of increase in soil temperature will stimulate the availability of nutrients for plants. Soil temperature also affects nutrient uptake by changing soil water viscosity and root nutrient transport, observed that at low soil temperature, nutrient uptake by plants reduce as a result of high soil water viscosity and low activity of root nutrient transport.

Root Growth

Increase in soil temperature improves root growth because of the increase in metabolic activity of root cells and the development of lateral. Low soil temperature results in reduced tissue nutrient concentrations and as such decreases root growth.

Color

Colour is an obvious characteristic of soil. It can provide a valuable insight into the soil environment. Thus it can be very important in assessment and classification.

The most influential colours in a well-drained soil are white, red, brown and black. White indicates the predominance of silica (quartz), or the presence of salts; red indicates the accumulation of iron oxide; and brown and black indicate the level and type of organic matter. A colour triangle can be used to show the names and relationships between the influential colours.

Color of Surface Soils

Colors associated with surface soils are dependent on the amount of organic matter found in them. Surface soils are placed in one of five color classes:

- Very dark: approximately 5 percent organic matter

- Dark: approximately 3.5 percent organic matter

- Moderately dark: approximately 2.5 percent organic matter

- Light: approximately 2 percent organic matter Very light—approximately 1.5 percent organic matter

Color of Subsoils

Subsoil colors are associated with natural drainage of the soils while the soils were forming. The level of moisture in a soil affected iron compounds that gave color to the subsoil. Subsoil colors are classified as bright, dull, or mottle colored.

Subsoil Classes

Bright-colored subsoil is characteristically brown, reddish brown, or yellowish brown. Good drainage is what gives subsoil a bright color. This is because the iron found in the soil has been oxidized. This can be compared to metal that oxidizes or rusts when both moisture and air are present. Rust has a bright or orange color.

Dull-colored subsoil is gray or olive gray. Poor drainage results in subsoil that is dull colored. This is because the iron found in the soil has not been subject to air or oxygen. The iron compounds do not oxidize, and therefore they leave a grayish color.

Mottle-colored subsoil consists of clumps of both bright and dull colors mixed together. Somewhat poor drainage of the subsoil leads to the mottled look. This is because the soil was saturated with moisture for certain periods. Some soil clumps take on a gray color because of the periods of poor drainage. Other soil clumps assume a bright color because the soil was comparatively dry during other periods.

Other Factors Affecting Soil Color

Besides organic matter and drainage, soil color is influenced by parent material, age, and slope of the land.

Parent Material

The color of a soil is associated with the kind of material from which it was formed. Soils developed from sand or light-colored rock will be lighter. Those developed from darker materials, such as peat or muck, will be darker.

Age

As soils age, much of the darker color is lost due to the weathering process. Weathering causes organic matter in the soils to break down.

Slope

Soil on the tops of hills is usually lighter in color than soil in depressions or on level ground. One reason for this is the darker topsoil is washed off the hills, leaving the lighter subsoil exposed. Another explanation is there tends to be greater moisture on lower land. This allows more abundant growth of plants in the lower areas, in turn providing more organic matter and a darker color to lower soil. Also, moisture in the low-lying soil slows the decay of the organic matter.

Figure: Years of wind and water erosion have changed the soil on the top of this hill.

Measurment of Soil Color

- Soil colour should be determined on moist surfaces of freshly broken (not sliced) soil samples.

- Like any other soil property, colour must always be observed throughout soil profile, paying special attention to the differences between soil horizons. Colour characteristics such as mottle size, percentage and contrast should be observed and recorded.

- A system that uses specially printed colour charts (Munsell Soil Colour Charts) gives an international standard. It divides colour into wavelength, lightness, and colour saturation.

Munsell Color System

Red, brown, yellow, yellowish-red, grayish-brown, and pale red are all good descriptive colors of soil, but not very exact. Just as paint stores have pages of color chips, soil scientists use a book of color chips that follow the Munsell System of Color Notation. The Munsell System allows for direct comparison of soils anywhere in the world. The system has three components: hue (a specific color), value (lightness and darkness), and chroma (color intensity) that are arranged in books of color chips. Soil is held next to the chips to find a visual match and assigned the corresponding Munsell notation. For example, a brown soil may be noted as: hue value/chroma (10YR 5/3). With a soil color book with Munsell notations, a science student or teacher can visually connect soil colors with natural environments of the area, and students can learn to read and record the color, scientifically. Soil color by Munsell notation is one of many standard methods used to describe soils for soil survey. Munsell color notations can be used to define an archeological site or to make comparisons in a criminal investigation. Even carpet manufacturers use Munsell soil colors to match carpet colors to local soils so that the carpet will not show the dirt (soil) tracked into the house.

Interpreting Soil Color

Color can be used as a clue to mineral content of a soil. Iron minerals, by far, provide the most and the greatest variety of pigments in earth and soil.

Relatively large crystals of goethite give the ubiquitous yellow pigment of aerobic soils. Smaller goethite crystals produce shades of brown. Hematite (Greek for blood-like) adds rich red tints. Large hematite crystals give a purplish-red color to geologic sediments that, in a soil, may be inherited from the geologic parent material. In general, goethite soil colors occur more frequently in temperate climates, and hematite colors are more prevalent in hot deserts and tropical climates.

Color - or lack of color - can also tell us something about the environment. Anaerobic environments occur when a soil has a high water table or water settles above an impermeable layer. In many soils, the water table rises in the rainy season. When standing water covers soil, any oxygen in the water is used rapidly, and then the aerobic bacteria go dormant. Anaerobic bacteria use ferric iron (Fe^{3+}) in goethite and hematite as an electron acceptor in their metabolism. In the process, iron is reduced to colorless, water-soluble ferrous iron (Fe^{2+}), which is returned to the soil. Other anaerobic bacteria use Mn^{4+} as an electron acceptor, which is reduced to colorless, soluble Mn^{2+}. The loss of pigment leaves gray colors of the underlying mineral. If water stays high for long periods, the entire zone turns gray.

When the water table edges down in the dry season, oxygen reenters. Soluble iron oxidizes into characteristic orange colored mottles of lepidocrocite (same formula as goethite but different crystal structure) on cracks in the soil. If the soil aerates rapidly, bright red mottles of ferrihydrite form in pores and on cracks. Usually ferrihydrite is not stable and, in time, alters to lepidocrocite.

Along seacoasts, tide waters saturate soils twice daily, bringing soluble sulfate anions. Anaerobic bacteria use the sulfate as an electron acceptor and release sulfide (S^{2-}) which combines with ferrous iron to precipitate black iron sulfide. A little hydrochloric acid (HCl) dropped on this black pigment quickly produces a rotten egg odor of hydrogen sulfide (H_2S) gas. Soils that release H_2S gas are called sulfidic soils. With time, iron sulfide alters to pyrite (FeS_2) and imparts a metallic bluish color. If sulfidic soils are drained and aerated, they quickly become very acid (pH 2.5 to 3.5), and a distinctive pale yellow pigment of jarosite forms. This is the mark of an acid sulfate soil that is quite corrosive and grows few plants.

Galuconitic green sands form in shallow ocean water near a coast. They become part of soils that form after sea level drops. White colors of uncoated calcite, dolomite, and gypsum are common in geologic materials and soils in arid climates. A little carbonate dissolves in water, moves downward, and precipitates in soft white bodies or harder nodules. It also accumulates in root pores as lacy, dendritic (tree-branch) patterns.

Resistivity

The measure of the resistance offered by the soil in the flow of electricity is called the soil resistivity. The resistivity of the soil depends on the various factors like soil composition, moisture, temperature, etc. Generally, the soil is not homogenous, and their resistivity varies with the depth. The soil having a low resistivity is good for designing the grounding system. The resistivity of the soil is measured in ohmmeter or ohm-centimeters.

The resistivity of the soil mainly depends on its temperature. When the temperature of the soil is more than $0°$, then its effect on soil resistivity is negligible. At $0°$ the water starts freezing and resistivity increases. The magnitude of the current also affects the resistivity of the soil. If the magnitude of current dissipated in the soil is high, it may cause significant drying of soil and increase its resistivity.

The resistivity of the soil varies with the depth. The lower layers of the soil have greater moisture content and lower resistivity. If the lower layer contains hard and rocky layers, then their resistivity may increase with the depth.

Measurement of Soil Resistivity

Wenner Soil Resistivity Testing and Other 4-Point Tests

The Wenner 4-point Method is by far the most used test method to measure the resistivity of soil. Other methods do exist, such as the General and Schlumberger methods, however they are infrequently used for grounding design applications and vary only slightly in how the probes are spaced when compared to the Wenner Method.

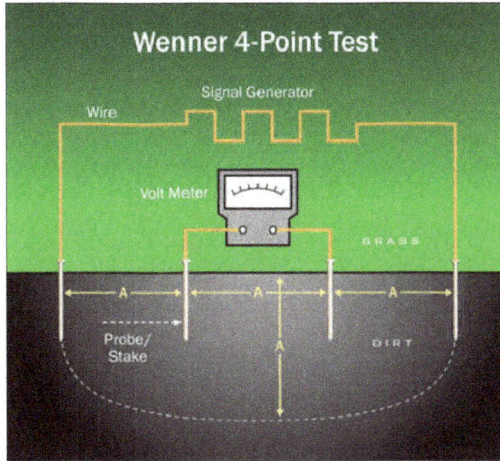

Wenner 4-Point Test

Electrical resistivity is the measurement of the specific resistance of a given material. It is expressed in ohm-meters and represents the resistance measured between two plates covering opposite sides of a 1 m cube. This soil resistivity test is commonly performed at raw land sites, during the design and planning of grounding systems specific to the tested site.

The soil resistivity test spaces four probes out at equal distances to approximate the depth of the soil to be tested. Typical spacings will be 1', 1.5', 2', 3', 4.5', 7', 10', etc., with each spacing increasing from the preceding one by a factor of approximately 1.5, up to a maximum spacing that is commensurate with the 1 to 3 times the maximum diagonal dimension of the grounding system being designed, resulting in a maximum distance between the outer current electrodes of 3 to 9 times the maximum diagonal dimension of the future grounding system. This is one "traverse" or set of measurements, and is typically repeated, albeit with shorter maximum spacings, several times around the location at right angles and diagonally to each other to ensure accurate readings.

The basic premise of the soil resistivity test is that probes spaced at 5' distance across the earth, will read 5' in depth. The same is true if you space the probes 40' across the earth, you get a weighted average soil resistance from 0' down to 40' in depth, and all points in between. This raw data is usually processed with computer software to determine the actual resistivity of the soil as a function of depth.

Conducting a Wenner 4-point (or four-pin) Soil Resistivity Test

The following describes how to take one "traverse" or set of measurements. As the "4-point" indicates, the test consists of 4 pins that must be inserted into the earth. The outer two pins are called the Current probes, C1 and C2. These are the probes that inject current into the earth. The inner two probes are the Potential probes, P1 and P2. These are the probes that take the actual soil resistance measurement.

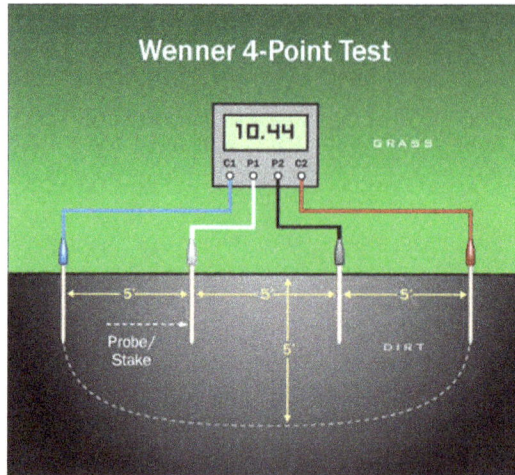

In the following Wenner 4-Point Test Setup diagram, a probe C1 is driven into the earth at the corner of the area to be measured. Probes P1, P2, & C2 are driven at 5', 10' & 15' respectively from rod C1 in a straight line to measure the soil resistivity from 0' to 5' in depth. C1 & C2 are the outer probes and P1 & P2 are the inner probes. At this point, a known current is applied across probes C1 & C2, while the resulting voltage is measured across P1 & P2. Ohm's law can then be applied to calculate the measured apparent resistance.

Probes C2, P1 & P2 can then be moved out to 10', 20' & 30' spacing to measure the resistance of the earth from 0' to 10' in depth. Continue moving the three probes (C2, P1 & P2) away from C1 at equal intervals to approximate the depth of the soil to be measured. Note that the performance of the electrode can be influenced by soil resistivities at depths that are considerably deeper than the depth of the electrode, particularly for extensive horizontal electrodes, such as water pipes, building foundations or grounding grids.

Soil Resistance Meters

There are basically two types of soil resistance meters: Low-Frequency and High-Frequency models. Both meter types can be used for 4-point & 3-point testing, and can even be used as standard (2-point) volt meter for measuring common soil resistivity.

Care should always be given when selecting a soil resistance meter, as the electronics involved in signal filtering are highly specialized. Electrically speaking, the earth can be a noisy place. Overhead power lines, electric substations, railroad tracks, various signal transmitters and many other sources contribute to signal noise found in any given location. Harmonics, 60 Hz background noise, and magnetic field coupling can distort the measurement signal, resulting in apparent soil resistivity readings that are larger by an order of magnitude, particularly with large spacings. Selecting equipment with electronic packages capable of discriminating between these signals is critical.

High-Frequency soil resistance meters typically use a pulses operating at 128 pulses per second, or other pulse rates except 60. These High-Frequency meters typically suffer from the inability to generate sufficient voltage to handle long traverses and generally should not be used for probe spacings greater than 100 feet. Furthermore, the High-Frequency signal flowing in the current lead induces a noise voltage in the potential leads, which cannot be completely filtered out: this noise becomes greater than the measured signal as the soil resistivity decreases and the pin spacing increases. High-Frequency meters are less expensive than their Low-Frequency counterparts, and are by far the most common meter used in soil resistivity testing.

Low-Frequency meters, which actually generate low frequency pulses (on the order of 0.5 to 2.0 seconds per pulse), are the preferred equipment for soil resistivity testing, as they do away with the induction problem from which the High-Frequency meters suffer. However they can be very expensive to purchase. Depending upon the equipment's maximum voltage, Low-Frequency meters can take readings with extremely large probe spacings and often many thousands of feet in distance. Typically, the electronics filtering packages offered in Low-Frequency meters are superior to those found in High-Frequency meters. Caution should be taken to select a reputable manufacturer.

Data Analysis

Once all the soil resistivity data is collected, the following formula can be applied to calculate the apparent soil resistivity in ohm-meters:

4-POINT DATA INTERPRETATION

$$\rho = 1.915\,\mathrm{AR}$$
$$\rho = 1.915(40)(4.5)$$

$$\rho_a = \frac{4\pi\,\mathrm{AR}}{1 + \dfrac{2\,\mathrm{A}}{\sqrt{\left(\mathrm{A}^2 + 4\,\mathrm{B}^2\right)}} - \dfrac{2\,\mathrm{A}}{\sqrt{\left(4\,\mathrm{A}^2 + 4\,\mathrm{B}^2\right)}}}$$

$\rho = $ Resistivity $\mathrm{B} = $ Depth of Probes

$\mathrm{A} = $ Spacing of Probes $\mathrm{R} = $ Resistance (reading from meter)

If $\mathrm{A} > 20\,\mathrm{B}$, then $\rho = 2\,\pi\,\mathrm{AR}$ $= 1.915\,\mathrm{AR}$

For example, if an apparent soil resistance of 4.5 ohms is at a 40-foot spacing, the soil resistivity in ohm-meters would be 344.7. Figure shows the entire soil resistivity formula in detail. One refers to "apparent" resistivity, because this does not correspond to the actual resistivity of the soil. This raw data must be interpreted by suitable methods in order to determine the actual soil resistivity.

Shallow Depth Readings

Shallow depth readings, as little as 6" in depth, are exceedingly important for most, if not all, grounding designs. As described above, the deeper soil resistivity readings are actually weighted averages of the soil resistivity from the earth surface down to depth, and include all the shallow resistance readings above it. The trick in developing the final soil model is to pull out the actual resistance of the soil at depth, and that requires "subtracting" the top layers from the deep readings. The following figure demonstrates how the shallowest readings impact deeper ones below it.

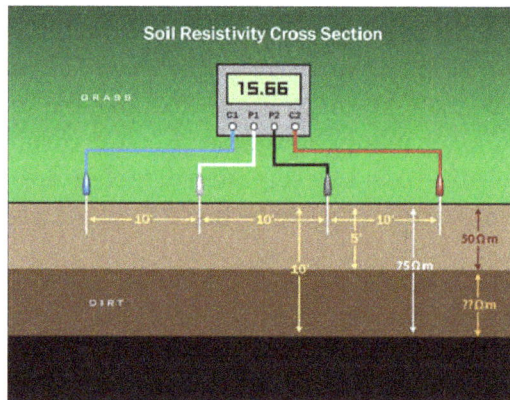

As you can see in the following diagram, if you have a 5' reading of 50 ohm-meters and a 10' reading of 75-ohmmeter soil, the actual soil resistivity from 5' to 10' might be 100 ohm-meters (the point here is to illustrate a concept: pre-computed curves or computer software are needed to properly interpret the data). The same follows true for larger pin spacings. The shallowest readings are used over and over again in determining the actual resistivity at depth.

Shallow depth readings of 6-inches, 1-foot, 1.5-feet, 2-feet and 2.5-feet are important for grounding design, because grounding conductors are typically buried at 1.5 to 2.5-feet below the surface of the earth. To accurately calculate how those conductors will perform at these depths shallow soil readings must be taken. These shallow readings become even more important when engineers calculate Ground Potential Rise, Touch Voltages and Step Voltages.

It is critical that the measurement probes and current probes be inserted into the earth to the proper depth for shallow soil resistivity readings. If the probes are driven too deep, then it can be difficult to resolve the resistivity of the shallow soil. A rule of thumb is that the penetration depth of the potential probes should be no more than 10% of the pin spacing, whereas the current probes must not be driven more than 30% of the pin spacing.

Deep Readings

Often, the type of meter used determines the maximum depth or spacing that can be read. A general guideline is that High-Frequency soil resistivity meters are good for no more than 100-feet pin spacings, particularly in low resistivity soils. For greater pin spacings, Low-Frequency soil resistivity meters are required. They can generate the required voltage needed to push the signal through the soil at deep distances and detect a weak signal, free of induced voltage from the current injection leads.

Effects of Soil Resistivity on Ground Electrode Resistance

Soil resistivity is the key factor that determines what the resistance of a grounding electrode will be, and to what depth it must be driven to obtain low ground resistance. The resistivity of the soil varies widely throughout the world and changes seasonally. Soil resistivity is determined largely by its content of electrolytes, which consist of moisture, minerals and dissolved salts. A dry soil has high resistivity if it contains no soluble salts.

Resistivity (approx), Ohm-centimeters			
Soil	Minimum	Average	Maximum
Ashes, cinders, brine, waste	590	2370	7000
Clay, shale, gumbo, loam	340	4060	16,300
Same, with varying proportions of sand and gravel	1020	15,800	135,000
Gravel, sand, stones with little clay or loam	59,000	94,000	458,000

Factors Affecting Soil Resistivity

Two samples of soil, when thoroughly dried, may in fact become very good insulators having a resistivity in excess of 109 ohm-centimeters. The resistivity of the soil sample is seen to change quite rapidly until approximately 20% or greater moisture content is reached.

Moisture content % by weight	Resistivity (Ohm-centimeters)	
	Top Soil	Sandy Loam
0	>109	>109
2.5	250,000	150,000
5	165,000	43,000
10	53,000	18,500
15	19,000	10,500
20	12,000	6300
30	6400	4200

The resistivity of the soil is also influenced by temperature. Figure shows the variation of the resistivity of sandy loam, containing 15.2% moisture, with temperature changes from 20° to -15° C. In this temperature range the resistivity is seen to vary from 7200 to 330,000 ohm-centimeters.

Temperature C	Temperature F	Resistivity (Ohm-centimeters)
20	68	7,200
10	50	9,900
0	32 (water)	13,800
0	32 (ice)	30,000
-5	23	79,000
-15	14	330,000

Because soil resistivity directly relates to moisture content and temperature, it is reasonable to assume that the resistance of any grounding system will vary throughout the different seasons of the year. Such variations are shown in Figure. Since both temperature and moisture content become more stable at greater distances below the surface of the earth, it follows that a grounding system, to be most effective at all times, should be constructed with the ground rod driven down a considerable distance below the surface of the earth. Best results are obtained if the ground rod reaches the water table.

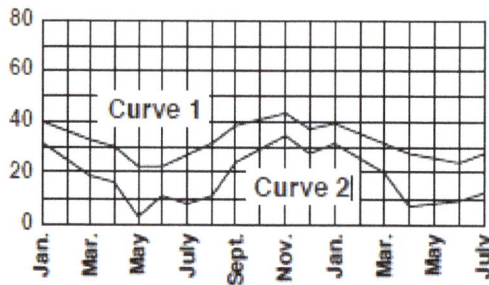

Seasonal variation of earth resistance with an electrode of 3/4-inch pipe in rather stony clay soil.
Depth of electrode in earth is 3 ft. for Curve 1, and 10 ft. for Curve 2

In some locations, the resistivity of the earth is so high that low-resistance grounding can be obtained only at considerable expense and with an elaborate grounding system. In such situations, it may be economical to use a ground rod system of limited size and to reduce the ground resistivity by periodically increasing the soluble chemical content of the soil. Figure shows the substantial reduction in resistivity of sandy loam brought about by an increase in chemical salt content.

The Effect of Salt* Content on the Resistivity of Soil (Sandy loam, Moisture content, 15% by weight, Temperature, 17°C)	
Added Salt (% By weight of moisture)	Resistivity (Ohm-centimeters)
0	10,700
0	1 1800
1	0 460
5	190
10	130
20	100

Chemically treated soil is also subject to considerable variation of resistivity with temperature changes. If salt treatment is employed, it is necessary to use ground rods, which will resist chemical corrosion.

The Effect of Temperature on the Resistivity of Soil Containing Salt (Sandy loam, 20% moisture, Salt 5% of weight of moisture)	
Temperature C	Resistivity (Ohm-centimeters)
20	110
10	142
1.00	190
-5	312
-13	1440

*Such as copper sulfate, sodium carbonate, and others

Salts must be EPA or local ordinance approved prior to use.

References

- Density-of-soil-meaning-and-types-1724: soilmanagementindia.com, Retrieved 17 March 2018

- Porosity-of-soil-meaning-and-factors-affecting-1720: soilmanagementindia.com, Retrieved 19 June 2018

- Soil-porosity-information, soil-fertilizers: gardeningknowhow.com, Retrieved 27 July 2018

- Soil-consistency, consistency-limits-of-soils-types-and-uses-13353: soilmanagementindia.com, Retrieved 07 July 2018

- Soil-resistivity: circuitglobe.com, Retrieved 10 April 2018

- What-is-soil-resistivity-testing: esgroundingsolutions.com, Retrieved 16 June 2018

3

Vadose Zone

The vadose zone is the unsaturated part of the subsurface of the Earth between the surface of the Earth and the water table. The aim of this chapter is to explore the varied aspects of vadose zone, such as movement of water in unsaturated and saturated zone, soil-water content, soil-water content measurement methods, heat flux through soil, etc. for an in-depth understanding.

The vadose zone (VZ) refers to the shallow zone of unsaturated porous media roughly between the land surface and the groundwater. Having varying thickness from a few centimeters in wetlands to several hundreds of meters in arid climates, the VZ is characterized by porous media (soil or fractured bedrock) partially saturated with a wetting fluid (e.g., water). The dynamics of water in this zone are intrinsically linked to the hydrologic cycle via partioning of water at the land surface and regulating the movement to and from the groundwater, thereby effectively governing interrelationships between precipitation, surface runoff, infiltration, groundwater recharge and evapotranspiration. As such, the VZ takes a central role in the critical zone describing the most heterogeneous and complex region for life on Earth that encompasses the region between the top of the vegetation canopy to the bottom of the groundwater aquifer in which rock, soil, water, air and living organisms regulate the natural habitat. As one central theme, water fundamentally frames environmental, hydrologic, socio-economic, and agricultural problems in this zone. Through the lens of water flow and distribution, the disciplines of soil physics and hydrology study the VZ.

The following pillar:

1. There is no panacea for VZ modelling approaches for all scales and interactions.

2. Water regulates much of the interactions with life in the VZ. Its status determines hydration of organisms, access to nutrients, exchange of metabolic products, as well as critically influences thermal and chemical properties of soils contributing to habitat formation.

3. The VZ is a thin layer constituting a natural resource that contributes to the resilience of ecosystems by acting as a low-pass filter for variability in weather, resource availability and toxicity (e.g., contaminants and solutes).

4. The Richards equation is the most common modeling approach for VZ water from the sample to the field scale.

5. Pore- and watershed-scale models utilize alternative modeling approaches with up and down-scaling bridging the scale gap or addressing a parameterization problem.

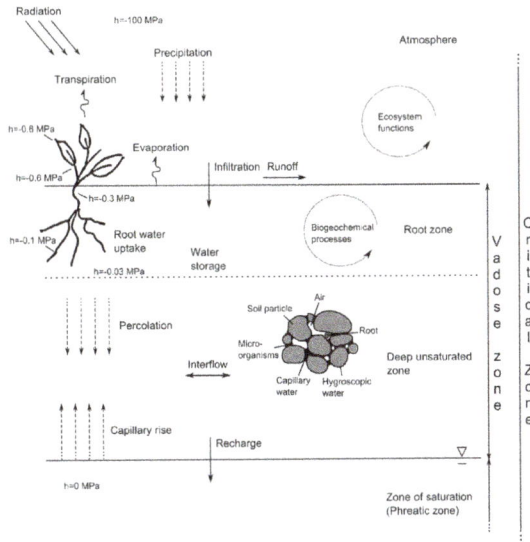

Figure: Fate and distribution of water in the vadose zone.

Characterized by porous media along with water and air which distribute according to energy potentials closely tied to pore sizes and shapes, various constituents and participants benefit from, and modify the VZ environment. For example, plants compete with the porous media for capillary water held in the pore space and remove water from the root zone driven by a gradient of soil-water potentials h which become increasingly more negative. With decreasing water contents in the VZ, the connectivity for gas-filled pathways that facilitate aeration increases. Similarly, thermal, chemical, pedological and biological processes are critically affected by water contents and fluxes. To facilitate optimal biogeochemical processes and support resilient ecosystem functions in the critical zone, some optimal balance (i.e., a critical window) of soil water, gas exchange and nutrient availability must be maintained in the VZ.

Using these five pillars we aim to make VZ processes and modeling accessible to non-VZ students with a strong bias towards water resources. Necessarily, we cannot provide a complete treatment of VZ processes and interactions nor a complete treatment of modeling approaches to facilitate a dialogue between disciplines, we have tried to reduce technical jargon and mathematic rigor in trying to explain what VZ modelers are doing.

Reason why VZ Processes are Inherently Interdisciplinary

The study of the VZ has long been central to agronomic, hydrologic, ecologic, and engineering sciences. Yet the VZ is a complex and dynamic system driven by natural processes and affected by human activities that defy disciplinary borders and place the VZ in a larger interdisciplinary context as the seminal crucible of terrestrial life.

The VZ is a diverse system composed of "constituents" (i.e., solids, liquids, solutes and gases) and "participants" (i.e., microbes, animals and plants) whose development and function is shaped by climate, parent material, topography, humans and time. Despite its relative thinness compared to the atmosphere, the lithosphere and even the phreatic zone (saturated aquifer), the VZ plays a central role in the terrestrial hydrologic cycle with its capacity to absorb, store, and transmit water among these spheres. The VZ also acts as a filter and buffer for dissolved or suspended compounds, thereby determining fate and transport in biogeochemical cycling closely linked to the hydrologic functions. These coupled processes in the VZ are central to mutual interactions involving water, air, soil, rock, organisms, and humans that regulate the natural habitat.

The partioning of energy and mass at the land surface as well as the distribution and fluxes within the VZ are determined by interactions of porous media solids with fluids (i.e., water and gases), microbial and vertebral participants, as well as plant roots. Water, because of its tendency to wet solid surface (i.e., forming liquid films, bridges and curved interfaces), largely controls the connectivity and tortuosity of pathways for the flow of water and gases, thereby mediating biogeochemical reactions dependent on some balance and composition of these phases. Therefore textural and structural porous media properties fundamentally determine whether incoming precipitation will generate overland flow, infiltrate, be made available to plant roots, or contribute to groundwater recharge. On the other hand, solar radiation absorbed by the land surface and plants drives evaporation, condensation, transpiration and biochemical reactions that exchange heat, water and gases between the VZ and the atmosphere.

Being at the nexus of water and energy fluxes, VZ processes are essential for water and ecosystem health where soil functions depend on a small degree of multitude soil organisms. This biodiversity and ecological functions are interrelated at local, regional and global scales. VZ resource pools and organism interactions alter dynamics of water and nutrient resource availability such that they alter the functional stability and ecological resilience. VZ water in particular constrains the resources for biota, which in turn control the spatial and temporal patterns of vegetation. Feedback loops such as these highlight the mutual impacts of VZ processes on the climate. Plants, above all, play a central role in theses feedbacks acting as the chief connector between atmosphere, below groun strata, and organisms in the exchange of greenhouse gases (i.e., water vapor, carbon dioxide, methane, and nitrous oxide). Carbon dioxide synthesized with VZ water and nutrients then forms the largest terrestrial organic carbon stock (second in size only to the oceans) which interacts strongly with the atmosphere, climate and land-use change to mediate the global greenhouse effect.

The complexity of interactions within and between VZ constituents and participants necessitate interdisciplinary study and modeling approaches. Consequently, many VZ models address fluxes, distribution and exchange of energy and mass at the boundaries and within the VZ, its interactions with plant roots and organisms as well as interactions, exclusions and accelerated transport of chemicals and colloids. However, because the study of the VZ has historically been disciplinary, simulation models tend to express linkages between constituents and participants explicitly. There is also a disjuncture between characteristic scales for these processes: whereas soil science deals with phenomena on the scale of a vertical profile or a restricted field, hydrology typically operates on a watershed level. Future advances in VZ models will benefit from advances in our understanding of fundamental physical, chemical and biological processes and interactions, but maybe even more so from synergistic applications from users outside of the traditional VZ sciences.

Metrics of VZ Water Status: How much is there and How Easy is it to Extract

Many important functions in the VZ such as heat and solute transport, biological processes, water supply to plants, evapotranspiration, groundwater recharge, and runoff generation are controlled by the state of water in the VZ. One of the tenants of VZ water is to describe its availability in addition to the amount. The availability of water recognizes that interactions with solids in the presence of solutes alter the energy required to forcibly move water in the VZ and determine the flow of water within the VZ. The availability is described as a specific energy potential H. It is recognized that water moves within the VZ according to gradients in this energy potential.

The water content most often provided as a volumetric water content $\theta_v [L^3\ L^{-3}]$ describes the volume fraction of water with respect to the bulk soil as $\theta_v = V_{water} / V_{soil}$. A wide variety of methods and sensors exist that make θ_v the most easily obtained VZ water status property. Nevertheless, given the immense variability in the VZ, the lack of spatially and temporally exhaustive water-content data remains one of the biggest limitations for furthering our understanding of this complex zone to date.

The water content θ_v alone is insufficient to characterize soil-water status. Like all other forms of matter, water flows from locations with high potential energy to locations of lower potential H energy in pursuit of an equilibrium state. The specific energy potential H (i.e., a potential per unit quantity of water) used to describe the energy potential of water in porous media is a measure relative to some reference state (free water at a standard atmosphere). H may be partitioned into multiple summative components: matric h (describing forces of interaction between solid and liquid), gravity h_z (position in a gravity field), osmotic (presence of solutes which tend to increase the surface tension), and pressure (height of free water above a point of interest, or pore water pressure due to restrictive horizons). To simplify, the total hydraulic potential is often considered to consist of only matric h and gravitational potential h_z. The following expression for the 1-dimensional vertical gradient in potentials at steady state can be written when potentials are expressed as energy per unit of weight (i.e., resulting in units of length [L]):

$$\left(\frac{dH}{dx}\right)=\left(\frac{dh}{dx_i}+\frac{dh_z}{dx_z}\right)=\left(\frac{dh}{dx_i}+1\right)$$

For horizontal flow, the gradient $dh_z/dz = 0$. Bittelli provides a recent review of measurement methods. The value of h ranges from zero, when the soil is water-saturated, to very low negative numbers when the soil is very dry (i.e., we need to invest work to retrieve water).

A Word on Models in General

In vadose zone physics, and science in general, we frequently wish to make inferences about physical parameters from measured data. The aim is to reconstruct the "real world" from a set of measurements. We call this the inverse problem. In the ideal case, an exact theory is known that describes how the data (describing state variables) should be transformed in order to reproduce the "real world." In reality, the theory is incomplete and not exact; we therefore have to make do with an approximating theory to the "real world." We call this the model. The true theory would be the model that describes reality. Models are always a simplified representation of the real world. However, we can use the model to simulate the measured data. This is called the forward problem. In many cases, the model is a continuous function of space and time variables, and has infinite degrees of freedom. But measured data are finite. Therefore, the data do not contain enough information to uniquely reconstruct the model (remember that the model may also not be a good representation of the "real world"). In addition, the measured data are averages over space and time, and worse, the data also contain some amount of noise. Forward and inverse problems are therefore plagued with two fundamental issues: non-uniqueness and error propagation.

A Strong Tradition of using the Richards Equation to Model VZ Water

The fundamental law for steady-state water flow in porous media is named after Henry Darcy whose seminal work described the flux density $j_w [L^3 \ L^{-2} \ T^{-1}]$ as the volumetric flow per cross sectional area due to a gradient in hydraulic potential mediated by the saturated hydraulic conductivity $K_s [L \ T^{-1}]$ as the factor of proportionality:

$$j_w = -K_s\left(\frac{dH}{dx}\right)$$

K_s lumps together attributes determining the ease of water movement in saturated soils like poresize, pore-shape and the tortuosity of water-filled pathways, as well as viscous properties. Equation ($j_w = -K_s\left(\frac{dH}{dx}\right)$) also encapsulates that water will flow from locations of high potential to locations of low potential. At unsaturated conditions, which define the VZ, Darcy's law remains applicable when introducing a hydraulic conductivity term $K(h) [L \ T^{-1}]$:

$$j_w = -K(h)\left(\frac{dH}{dx}\right)$$

Equation ($j_w = -K(h)\left(\dfrac{dH}{dx}\right)$) is referred to as the Buckingham-Darcy law. In it, the factor of proportionality, K(h), is reduced compared to the saturated case and is a highly non-linear function of soil-water content and potential. The reduction can be thought of as a reduction in participating water volumes to the flow with increasing air-filled pore spaces. This reduction is highly non-linear, because of the increasing tortuosity with desaturation of the remaining water-filled connected pathways.

The general case of unsteady and unsaturated flow is a highly dynamic phenomenon which may be represented by combining Equation ($j_w = -K(h)\left(\dfrac{dH}{dx}\right)$) with continuity:

$$\frac{\partial \theta}{\partial t} = -\frac{\partial j_{wi}}{\partial x_i} - S$$

to yield the Richards (1931) equation:

$$\frac{\partial \theta}{\partial t} = -\frac{\partial}{\partial x_i}\left(K(h)\frac{\partial}{\partial x_i}(x_i + z)\right) - S$$

Where S [L⁰ T⁻¹] is a sink/source term (e.g., plant water uptake). Equation ($\frac{\partial \theta}{\partial t} = -\frac{\partial}{\partial x_i}\left(K(h)\frac{\partial}{\partial x_i}(x_i + z)\right) - S$) is a non-linear second-order partial differential equation (PDE) that describes the general flow of water in variably-saturated.

A major drawback of Equation ($\frac{\partial \theta}{\partial t} = -\frac{\partial}{\partial x_i}\left(K(h)\frac{\partial}{\partial x_i}(x_i + z)\right) - S$) is the dependence on both θ and h. Obtaining solutions to this equation therefore requires constitutive relations to describe the interdependence of h(θ), or to eliminate either θ or h from Equation $\frac{\partial \theta}{\partial t} = -\frac{\partial}{\partial x_i}\left(K(h)\frac{\partial}{\partial x_i}(x_i + z)\right) - S$. Like all other second-order PDEs, solutions to Equation ($\frac{\partial \theta}{\partial t} = -\frac{\partial}{\partial x_i}\left(K(h)\frac{\partial}{\partial x_i}(x_i + z)\right) - S$) require the specification of two

additional pieces of information that take the form of initial and boundary conditions. Initial conditions are specified as a complete description of the system in terms of parameters such as θ or h at some initial time (i.e., t = 0). Boundary conditions are specified as environmental conditions at the boundaries of the system. For example, a

1-dimensional vertical model describing infiltration of water into a soil profile may be set up by specifying environmental conditions at the soil surface and some depth z in terms of water content θ (Dirichlet condition), water content gradient δθ/δz (Neumann condition) or a linear combination of the two (Cauchy condition). The boundary conditions may change with time to describe effects such as changes in rate of precipitation or a fluctuating water table.

Analytical solutions to Equation ($\frac{\partial \theta}{\partial t} = -\frac{\partial}{\partial x_i}\left(K(h)\frac{\partial}{\partial x_i}(x_i + z)\right) - S$) are available only for a limited number of applications and are often obtained by making simplifying assumptions, and using approximate parameterizations, as well as initial and boundary conditions. With today's computing power, numerical solutions using finite-element or -difference methods provide convenient brute force methods. In addition, often one is interested in coupled flow and transport in soils which commonly involves setting up several general flow models and solving them iteratively. Luckily, several modeling packages are available to numerically solve these kinds of problems including, for example, Hydrus, TOUGH, MODFLOW, SWAP as well as domain spreading models such as SHAW and more general PDE solvers like COMSOL Multiphysics. However, a solution will only be as good as the input parameterizations as well as assumptions and limitations inherent in the models. Common assumptions of VZ models include steady-state, equilibrium conditions, homogeneity, ergodicity and isotropy.

Constitutive Relations Needed for Parameterizing the Richards Equation

Describing and predicting VZ-water is key to understanding interdisciplinary interactions. The movement and distribution of water depends on the hydraulic properties of the soil-water-air continuum, its intricate pore space, arrangement, surface area, and phase distribution. Together, these effects can be described through the soil water retention characteristic, and the hydraulic conductivity functions.

Soil-Water Retention Characteristic

$\theta(h)$ is primarily determined by the pore-size distribution and the pore shapes - it is thus a unique fingerprint for complex soils. Soils with wide pore-size distributions (e.g., loam) tend to retain more water at more negative potentials than soils with narrow pore-size distributions (i.e., sand) which tend to drain more rapidly. For a review on the theory. A common conceptual model for matric potentials and θ(h) in soils is capillary rise and the simplification of soil-pore space into a bundle of capillaries. The capillary rise model predicts the matric potential (as energy per unit weight) as a function of apparent pore radii:

$$h = \frac{2\sigma \cos \gamma}{\rho_w gr}$$

where σ [$M\ L^0\ T^{-2}$] is the surface tension, γ[-] is the contact angle, ρ_w [$M\ L^{-3}$] is the density of water, g [$L\ T^{-2}$] the acceleration due to gravity, and r [L] the apparent pore radius. Capillary rise is higher in soils with small pores because of a more negative matric potential. One can use capillary rise to visualize $\theta(h)$ by observing the height of rise against gravity and determining the average water content of the soil at different elevations. Close to the water surface, large and small pores are water-filled, but the higher up (i.e., at lower potentials) one looks, only smaller pores remain water-filled.

For modeling and analysis it is beneficial to represent the $\theta(h)$ as a continuous parametric function. Commonly used parametric models are the van Genuchten and Brooks and Corey (1964) relationships. The van Genuchten model is given as:

$$\Theta(h) = \frac{\theta_v - \theta_r}{\theta_s - \theta_r} - \left\{\left[1 + \left(a|h|\right)^n\right]^{-m}\right\} \quad for\ h < 0$$

$$1 \qquad\qquad for\ h \geq 0$$

where a [L^{-1}], n [-] and m [-] are empirical fitting parameters. Θ [-] is the degree of saturation, θ_r and θ_s[$L^3\ L^{-3}$] are the residual and saturated water contents, respectively. These unknown parameters can be obtained by fitting the model to measured data pairs of volumetric water content θ_v and matric potential h using, for example, freely accessible RETC software. Multiple effects related to advancing and receding water in porous media result in $\theta(h)$ being hysteretic, i.e., drying and wetting processes have distinct characteristics. This hysteretic nature adds complexity to *VZ* modeling efforts. To further complicate matters, $\theta(h)$ may also be dynamic (e.g., in shrink swell soils, with root advance).

Unsaturated Hydraulic Conductivity Function

In saturated soils, *K(h = 0)* equals the saturated hydraulic conductivity *Ks* which constitutes the largest value for maximally conductive pathways with complete water-saturation. *K(h)* then describes the reduction in hydraulic conductivity with decreasing water content (reduction in cross sectional area for flow and participating flow pathways; increasing solid-liquid interactions and tortuosity). Because unsaturated conductivities are more difficult to measure, *K(h)* is most frequently modeled using a known *Ks* and some correlated soil properties to predict the shape of the decay function. One example is to use the $\theta(h)$ shape parameters to predict the shape of the *K(h)* function as suggested by Mualem.

Current direct determinations in the laboratory or field for $\theta(h)$ are described in Klute and Page and Rawls, and for *K(h)* in Dane. Alternatively, one may use parameter optimization to indirectly estimate $\theta(h)$ and *K(h)* from transient flow data by modelling the flow process and minimizing some objective function describing the differences between the measured and predicted flow variables. However, both direct and indirect approaches to parameter estimation are often difficult and time-consuming, resulting in a general lack of information to properly parameterize the spatial and temporal variability in the VZ. Methods to partially overcome this limitation are the use of hydraulic-parameter databases or pedotransfer functions that predict hydraulic parameters based on other more easily measured porous-media properties.

Scale Issues in VZ Modeling

Water flow in the VZ occurs at different spatial scales ranging from the interface of solids and fluids, clusters of pores, pedon, to the field and watershed scale with variations in the dominant hydrologic flow processes. This extremely large spread in scales, coupled with the inherent complexity and non-linearity of flow in heterogeneous soils with high spatial variability (often the vertical variability is much higher than the horizontal—this anisotropy is often exploited when we use 1-dimensional models) complicates VZ modeling. A rough categorization of modelling approaches with scale may be given as:

1. At the pore scale, interface shapes and energy potentials as determined by pore size, shape and fluid are recognized as modeling approaches using the Hagen–Poiseuille or Navier–Stokes equations. These approaches are based largely on energy conservation. Another approach is the use of percolation theory.

2. At the pedon and field scale, VZ processes are often modeled using the Richards equation based on energy and mass conservation requiring the formulation and quantification of constitutive relations.

3. At the regional scale, VZ processes are recognized for controlling both short-term dynamics in watershed hydrology and long-term water balances of hydrologic basins. This approach, primarily motivated by mass conservation, is advantageous because relationships linking mass to energy states are difficult to obtain at this scale.

The boundaries between the scales are rather fluid, and processes at large scales may be predicted using small scale models and vice versa. With advances in computing power, soil physicists are increasingly upscaling their models to yield predictions of flow processes based on small-scale parameters that capture large-scale behaviour of heterogeneous VZs in an average sense, while hydrologists are downscaling in an attempt to predict small-scale processes based on averaged large-scale parameters.

Other issues of scale are phenomena such as rapid transport through macropores, preferential flow due to the spatial variability of hydraulic properties, or the instability of wetting fronts. Provides a review of modeling approaches for describing nonequilibrium and preferential flow.

Phreatic Zone

The zone of saturation is the ground immediately below the water table. The pores and fractures in soil and rocks are saturated with water.

The zone of saturation is less corrosive than the unsaturated zone above the water table. The moisture content in the region is at one extreme while the other extreme is in the dry soil. Maximum corrosion occurs at the intermediate of the two extremes of soil-moisture content.

The zone of saturation is also known as the phreatic zone.

The zone of saturation can be found anywhere between a few feet to over thousands of feet below the surface. It holds most of the world's fresh drinking water, which can be accessed from springs, rivers and wells. This water is sometimes polluted by human activity like the use of fertilizers and pesticides, septic tanks and landfills.

The size and depth of this region fluctuates as seasons change, and its levels depend on whether it is a dry or wet period, as well as other factors such as the drawing of water from wells and springs and other human activities.

The reason for the low corrosive atmosphere in the region is the low concentration of oxygen in the soil's moisture content. The little oxygen is not sufficient affect the metal surface, which is a prerequisite for corrosion to occur.

Other factors that may influence corrosion in the zone of saturation are the presence of dissolved substances like the chloride ions, sulphates and other aggressive substances.

Movement of Water in Unsaturated and Saturated Zone

The unsaturated zone is the region through which water, together with pollutant carried by the water, must pass to reach the groundwater. Therefore various processes occurring within the unsaturated zone play a major role in determining both the quality and quantity of water recharging into the groundwater. A quantitative study of water flow and contaminant transport in the unsaturated zone is a key factor in the improvement and protection of the quality of groundwater supplies.

Numerous simulation models for water flow and solute transport in the unsaturated zone are now being used increasingly for numerous applications in both research and management. Modelling techniques vary from straightforward analytical or semi-analytical methods to sophisticated numerical codes. Although analytical and semi-analytical methods remain widely used for certain applications, the growing power of personal computers along with the progression of more precise and numerically stable solution techniques have inspired significantly broader usage of numerical codes in recent years. The extensive utilization of numerical models is additionally greatly improved by their availability in both the commercial and public domains, and by the advancement of innovative graphics-based interfaces which significantly simplify their usage.

Mathematical Equations of Water and Transport in Unsaturated Soils

Analytical, semi-analytical, and numerical models are usually based on the following three governing equations for water flow, solute transport, and heat movement, respectively:

$$\frac{\partial \theta(h)}{\partial t} = \frac{\partial}{\partial z}\left[K(h)\left(\frac{\partial h}{\partial z}+1\right)\right] - S$$

$$\frac{\partial \theta Rc}{\partial t} = \frac{\partial}{\partial z}\left[\theta D\left(\frac{\partial c}{\partial z}-qc\right)\right] - \phi$$

$$\frac{\partial c(\theta)T}{\partial t} = \frac{\partial}{\partial z}\left[\lambda(\theta)\left(\frac{\partial T}{\partial z}\right)-C_w qT\right]$$

Suitable simplifications (mostly for analytical approaches) or extensions thereof (e.g. for two- and three-dimensional systems) are also employed. In equation ($\frac{\partial \theta(h)}{\partial t} = \frac{\partial}{\partial z}\left[K(h)\left(\frac{\partial h}{\partial z}+1\right)\right] - S$), frequently known as the Richards equation, h is the pressure head, z is the vertical coordinate positive upwards, θ is the water content, t is time, S is a sink term representing root water uptake or some other sources or sinks, and the hydraulic conductivity function (unsaturated) $K(h)$ is, often given as the product of the relative hydraulic conductivity, K_r, and the saturated hydraulic conductivity, K_s. In equation $\frac{\partial \theta Rc}{\partial t} = \frac{\partial}{\partial z}\left[\theta D\left(\frac{\partial c}{\partial z}-qc\right)\right] - \phi$, called the convection-dispersion equation (CDE), c is the solution concentration, D is the dispersion coefficient accounting for hydrodynamic dispersion and molecular diffusion, the retardation factor (R) that accounts for adsorption, the volumetric fluid flux density (q), and Φ is a sink/source term that accounts for various zero- and first-order or other reactions. In equation (3), T is temperature, λ is the apparent thermal conductivity, and C and C_w are the volumetric heat capacities of the soil and the liquid phase, respectively.

Solutions of the Richards equation ($\frac{\partial \theta(h)}{\partial t} = \frac{\partial}{\partial z}\left[K(h)\left(\frac{\partial h}{\partial z}+1\right)\right] - S$) require knowledge of the unsaturated soil hydraulic functions, that is, the soil water retention curve, $\theta(h)$, describing the relationship between the water content θ and the pressure head h, and the hydraulic conductivity function (unsaturated), $K(h)$, defining the hydraulic conductivity K as a function of h or θ. Under certain conditions (i.e. for linear sorption, a concentration-independent sink term Φ, and a steady flow field), equations $\frac{\partial \theta Rc}{\partial t} = \frac{\partial}{\partial z}\left[\theta D\left(\frac{\partial c}{\partial z}-qc\right)\right] - \phi$ and $\frac{\partial c(\theta)T}{\partial t} = \frac{\partial}{\partial z}\left[\lambda(\theta)\left(\frac{\partial T}{\partial z}\right)-C_w qT\right]$ are linear equations. But equation $\frac{\partial \theta(h)}{\partial t} = \frac{\partial}{\partial z}\left[K(h)\left(\frac{\partial h}{\partial z}+1\right)\right] - S$ is often very nonlinear due to the nonlinearity of the soil hydraulic properties. Consequently, numerous analytical solutions were derived previously for equations $\dfrac{\partial \theta Rc}{\partial t} = \dfrac{\partial}{\partial z}\left[\theta D\left(\dfrac{\partial c}{\partial z} - qc\right)\right] - \phi$ and $\dfrac{\partial c(\theta)T}{\partial t} = \dfrac{\partial}{\partial z}\left[\lambda(\theta)\left(\dfrac{\partial T}{\partial z}\right) - C_w qT\right]$ and these analytical solutions are still popular for evaluating solute and heat transport under steady-state conditions. Although a large number of analytical solutions of $\dfrac{\partial \theta(h)}{\partial t} = \dfrac{\partial}{\partial z}\left[K(h)\left(\dfrac{\partial h}{\partial z} + 1\right)\right] - S$ exist, they can generally be applied only to drastically simplified problems. Most of the applications of water flow in the vadose zone demand a numerical solution of the Richards equation.

Required Input Data

Simulation of water dynamics in the unsaturated zones needs input data regarding the model parameters, the geometry of the system, the boundary conditions and when simulating transient flow, initial conditions also. With geometry parameters, the dimensions of the problem domain are defined. With the physical parameters, the physical properties of the system under consideration are described. In relation to the unsaturated zone, it concerns $h(\theta)$ (soil water characteristic), and $K(\theta)$ (hydraulic conductivity).

For an appropriate explanation of the unsaturated flow, a proper description of the two hydraulic functions, $h(\theta)$ and $K(\theta)$, is important. $K(\theta)$, the hydraulic conductivity, decreases significantly when the moisture content (θ) decreases from saturation. The experimental approach to measure $K(\theta)$ at different moisture contents is fairly complicated and not too trustworthy. Alternate methods were therefore developed to determine the $K(\theta)$ function from more conveniently measurable characterizing properties of the soil. In many studies, the hydraulic conductivity of the unsaturated soil is defined as product of a non-linear function of the effective water saturation, together with hydraulic conductivity at saturation. The relationship is shown by

$$K(\theta) = K_s\left(\frac{\theta - \theta_r}{\theta_s - \theta_r}\right)^n$$

where,

K_s = hydraulic conductivity at saturation;

θ_s = saturated water content; and

θ_r = residual water content.

The value of n is found to be 3.5 for coarse textured soils. n will vary with soil type. In literature, established empirical correlation between n and soil characteristic is available. The relationship between the soil water pressure head $h(\theta)$ and moisture content θ, termed as the soil moisture characteristic or water retention curve, is normally determined by the textural and the structural composition of the soil. Also, the organic matter content may have an influence on the relationship. A characteristic feature of the water retention curve is that suction head ($-h$) decreases fairly rapidly with increasing moisture content. Hysteresis effects might emerge, and rather than being a single-valued relationship, the h-θ relation includes a group of curves. The actual curve will have to be determined from the history of wetting and drying.

When root water uptake is also modelled, the parameters describing the relation between root water uptake together with soil water status has to be supplied, along with crop specifications. If a functional flux-head relationship is employed as lower boundary condition, the parameters describing the interaction between surface water and groundwater and, if required, the vertical resistance of low permeable layers must be provided.

The number and kind of parameters necessary for modelling flow as well as transport processes in soils are dependent on the kind of model selected. These parameters can be categorized as control parameters (controlling the operation of the computer code), discretization data (grid and time stepping), and material parameters. The material parameters can be grouped in seven sets – static soil properties, water transport and retention functions, time-dependent parameters, basic chemical properties, contaminant source characteristics, soil adsorption parameters, and tortuosity functions. Table 1 lists many of the relevant material model parameters.

Table: Selected material parameters for flow and transport modelling

Model Parameters		
Static Soil Properties	Flow and Transport Variables and Properties	Basic Chemical Properties
Porosity		Molecular Weight
Bulk Density	Saturated Hydraulic Conductivity	Vapour Pressure
Particle Size	Saturated Water Content	Water Solubility
Specific Surface Area	Moisture Retention Function	Henry's Constant
Organic Carbon Content	Hydraulic Conductivity Function	Vapour Diffusion Coeff. in air
Cation Exchange Capacity	Dispersion Coefficient	Liquid Diffusion Coeff. in water
pH		Half-life or decay Rate
Soil Temperature		Hydrolysis Rate (s)
Time Dependent Parameters	Contaminant Source Characteristics	
Water Content	Soil Adsorption Parameters	
Water Flux	Distribution Coefficient	
Infiltration Rate	Isotherm Parameters	
Evaporation Rate	Organic Carbon Partition Coefficient	
Solute Concentration	Tortuosity Functions	
Solute Flux	Vapour Diffusion Tortuosity	
Solute velocity	Liquid Diffusion Tortuosity	
Air Entry Pressure Head		
Volatization Flux		

Modelling of Unsaturated Flow

Analytical solutions, if available, offer a greater understanding of the physics behind the transport phenomena and are computationally efficient and simple to use. However, analytical approaches are for the most part limited to situations of simple geometry domains, linear governing equations, and homogeneous systems. Along with efficient numerical methods and rapidly updated computer hardware, a large number of numerical models have been developed. However, the numerical technique cannot replace the analytical approaches completely, since numerical methods themselves need verification against analytical solutions because of discretization errors and convergence and stability problems that may be especially troublesome for advection-dominated and nonlinear adsorption processes.

Analytical solutions to the Richards equation for unsaturated flow under various boundary and initial conditions are difficult to obtain because of the nonlinearity in soil hydraulic parameters. This difficulty is exaggerated in the case where the soil is heterogeneous. Usually, one has to depend on numerical methods for predicting moisture movement in unsaturated soils, even for soils that are homogeneous. However, numerical approaches often suffer from convergence and mass balance problems.

Analytical solutions are only applicable to highly simplified systems and are not well suited for situations normally encountered in the field. Originally, finite difference techniques were primarily formulated to predict unsaturated flow solely; however finite element solution methods also were introduced later on. The nonlinearity of Richards equation is usually solved using an iterative procedure such as Newton or Picard methods. Perhaps the most important advantage of finite element techniques over standard finite difference methods is the ability to describe irregular system boundaries in simulations more accurately, as well as easily including non-homogeneous medium properties. For one-dimensional simulations, finite difference methods are just as good as finite element schemes. However, several authors suggest that finite element methods lead to more stable and accurate solutions, thus permitting larger time steps and/or coarser grid systems, and hence leading to computationally more efficient numerical schemes.

To numerically solve coupled systems of equations, the solution process requires some manipulation at each time step so that the dependence of one equation on the solution of the other is dealt with accurately. One way to overcome this is to use a fully implicit approach to solve the equations simultaneously. Any nonlinearity of the generated system can be handled by Newton's method. The implicit nature of this scheme allows for larger time steps in the simulation to find stable solutions as compared to the time steps for explicit schemes. An alternative to fully implicit scheme is to apply the mixed implicit-explicit approach. Yet somehow, the explicit portion of the scheme implies that this algorithm is now dependent upon a stability constraint which significantly limits the size of the time step and opens up numerical artifacts.

For heterogeneous soils that contain macro-pores, a different modelling approach is needed, as the presence of macro-pores in these soils may form a separate network for water flow. The common approach is to introduce two-region flow domain, one for macro-pores and the other for the soil matrix. In each flow domain, hydraulic conductivity is given independently. One more strategy would be to think of the heterogeneous soil as a group of stream tubes. It is assumed that there is no exchange of water between these tubes and that within each tube, the hydraulic conductivity is defined, but varies between tubes.

Modelling of Solute Transport

Movement of dissolved solutes in soils is often defined by the advection-dispersion equation. Analytical solutions have been derived for various boundary and initial conditions. Although these solutions are obtained for limited specified conditions, they have numerous applications like the verification of computer codes, prediction of solute movement for large times or distances where the use of numerical models become impractical, and the determination of transport parameters from soil column tests. The majority of analytical solutions pertain to semi-infinite and infinite media. Solutions are obtained by a variety of mathematical techniques including Green's functions, separation of variables, characteristics method, Laplace transforms and Fourier transforms.

Prediction of solute migration under field situations requires the concurrent solution of the solute transport and unsaturated flow equations. First approximations involve or assume steady flow and constant water contents. Because of the natural complexity of unsaturated flow, strategies for predicting solute transport have depended mostly on the finite difference or finite element approximations of the governing equations. Given that the equations for advection-diffusion transport usually do not have closed form analytical solutions, it is crucial that the numerical approximations be correct. When diffusion dominates the physical process, regular finite difference and finite element methods function effectively in solving these equations. But if advection is the dominant process, these equations could display numerous numerical problems. In fact, any standard finite difference or finite element method will produce a solution, which exhibits non-physical oscillations.

One of the distinctive features of the porous media on the field scale is the spatial heterogeneity of transport properties. These features have a distinct effect on the spatial distribution of contaminant concentration, as has been observed in field experiments and demonstrated by simulation of contaminant transport in unsaturated, heterogeneous soil. Description of the mixing process due to spatial variability of the unsaturated hydraulic conductivity has been advanced with the development of numerical solutions, which assume spatially variable soil properties; stochastic models; and stochastic stream tube models, which decompose the field into a set of independent vertical soil columns.

Unsaturated Zone Modelling Software

Most of the early models developed for studying processes in the near-surface environment focused mainly on variably saturated water flow. They were used primarily in agricultural research for the purpose of optimizing moisture conditions to increase crop production. This focus has increasingly shifted to environmental research, with the primary concern now being the subsurface fate and transport of various agricultural and other contaminants. While the earlier models solved the governing equations

$$\frac{\partial \theta(h)}{\partial t} = \frac{\partial}{\partial z}\left[K(h)\left(\frac{\partial h}{\partial z}+1\right)\right] - S \text{ through } \frac{\partial c(\theta)T}{\partial t} = \frac{\partial}{\partial z}\left[\lambda(\theta)\left(\frac{\partial T}{\partial z}\right) - C_w qT\right] \text{ for relative-}$$

ly simplified system-independent boundary conditions (i.e. specific pressure heads or fluxes, together with free water drainage), models developed recently could handle even more intricate system-dependent boundary conditions analyzing energy balances and surface flow and accounting for the concurrent movement of heat, vapour and water.

Movement of Water in Saturated Zone

The water that infiltrates through the unsaturated soil layers and move vertically ultimately reaches the saturated zone and raises the water table. Since it increases the quantity of in the saturated zone, it is also termed as 'recharge' of the ground water.

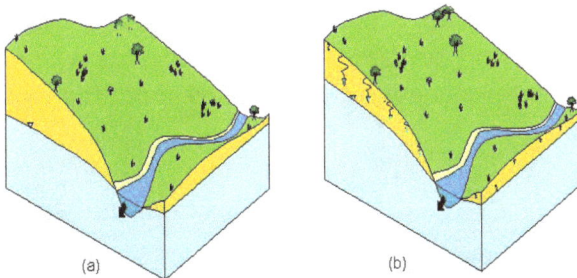

Figure: Variation of ground water table: (a) Before infiltration; (b) During/soon after infiltration

It may be observed from figure that both before any infiltration took place, there existed a gradient of the water table which showed a small gradient towards the river. However, the rise of the water table after the recharge due to infiltrating water is not uniform and thus the gradient of the water table after recharge is more than that before recharge. This has a direct bearing on the amount of ground water flow, which is proportional to the gradient. Based on actual observation or on mathematical analyses, we may draw lines of equal hydraulic head (the equipotential) within the saturated zone, as shown in figure. We may also draw the flow lines, which would be perpendicular to the equipotential lines. The flow lines, indicating the general direction of flow within the saturated soil zone is also drawn in the figure.

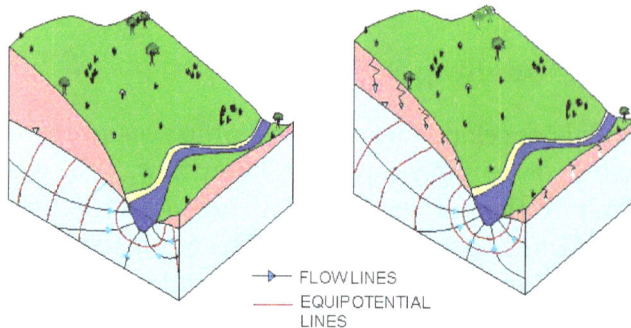

FLOW LINES
EQUIPOTENTIAL LINES

Figure: Equipotential lines and flow lines of ground water movements: (a) Before infiltration, with a low water table (b) During/soon after infiltration, with a raised water table

The rate of movement of the ground water, of course, varies with the material through which it is flowing since actually the flow is taking place through the voids which is different for different materials. The term hydraulic conductivity of a porous medium is used to indicate the ease with which water can flow through it. It is defined as the discharge taking place through a flow tube (which may be thought of as a short pipe along a flow line) per unit area of the tube under the influence of unit hydraulic gradient (which is the difference of potential heads in unit distance along the flow line). Hydraulic conductivity is generally denoted by 'K' and if the porous material is homogeneous, then K is also likely to be the same in any direction. However, in nature, the soil layers are often formed in layers resulting in the hydraulic conductivity varying between different directions. Even porous bed rock, which is usually fractured rock, may not be fractured to the same extent in all directions. As a result, in many natural flows the flow is more in some preferential direction. This type of conducting media is referred to as being heterogeneous and the corresponding hydraulic conductivity is said to be anisotropic.

Soil-Water Content

The moisture content of soil also referred to as water content is an in dicator of the amount of water present in soil. By definition, moisture content is the ratio of the mass of water in a sample to the mass of solids in the sample, expressed as a percentage. In equation form,

$$w = \frac{M_w}{M_s} \times 100$$

where:

> w = moisture content of soil expressed as a percentage
>
> M_w = mass of water in soil sample (i.e., initial mass of moist soil minus mass of oven-dried soil)
>
> M_s = mass of soil solids in sample (i.e., the soil's "oven-dried mass")

M_w and M_s may be expressed in any units of mass, but both should be expressed in the same unit.

It might be noted that the moisture content could be mistakenly de fined as the ratio of mass of water to total mass of moist soil rather than to the mass of oven-dried soil. Because the total mass of moist soil is the sum of the mass of water and oven-dried soil, this incorrect definition would give a fraction in which both numerator and de-nominator vary (but not in the same proportion) according to the amount of moisture present. Such a definition would be undesirable, because moisture content would then be based on a varying quantity of moist mass of soil rather than a constant quantity of oven-dried soil. Stated another way, with the incorrect definition, the moisture content would not be directly proportional to the mass of water present. With the correct definition given by Eq. $w = \dfrac{M_w}{M_s} \times 100$, moisture content is directly proportional to the mass of water present. This characteristic makes moisture content, as defined by Eq. $w = \dfrac{M_w}{M_s} \times 100$, one of the most useful and important soil parameters.

Effect of Soil Type

Soil moisture content is very dependent on soil type. A saturated coarse, sandy soil can hold far less water than saturated heavy silty clay. Sand has large particles which take up a lot of physical space. Also, as sand particles do not bind water, a lot of water will drain out of the sand due to gravity before field capacity is reached. For these two reasons, sand has a much lower maximum and minimum water content than a clay soil does.

What this means for you is that if you are monitoring a number of sites across your property, the soil moisture content values should not be compared between paddocks with different types of soil. Reading 20% moisture in one paddock and 20% in another paddock does not mean that the plants will be equally happy.

In sand, plants in 20% moisture will be very happy as sand readily releases its moisture and the most sand can hold is around 30%.

However, in clay, a plant in 20% moisture will be extremely stressed. Clay soils often have a maximum moisture reading of 50% or more, so 20% is very dry. The clay parti-cles also bind water to themselves and at low moisture contents like 20%, the clay will not give the water up for the roots to use.

It is therefore important, for each site need to be considered individually.

Importance of Soil Water

1. Soil water serves as a solvent and carrier of food nutrients for plant growth.

2. Yield of crop is more often determined by the amount of water available rather than the deficiency of other food nutrients.

3. Soil water acts as a nutrient itself.

4. Soil water regulates soil temperature.

5. Soil forming processes and weathering depend on water.

6. Microorganisms require water for their metabolic activities.

7. Soil water helps in chemical and biological activities of soil.

8. It is a principal constituent of the growing plant.

9. Water is essential for photosynthesis.

Retention of Water by Soil: The soils hold water (moisture) due to their colloidal properties and aggregation qualities. The water is held on the surface of the colloids and other particles and in the pores. The forces responsible for retention of water in the soil after the drainage has stopped are due to surface tension and surface attraction and are called surface moisture tension. This refers to the energy concept in moisture retention relationships. The force with which water is held is also termed as suction.

Soil-Water Content Measurement Methods

Soil water content can be measured directly or indirectly. In the first case, the amount of water is directly measured, for instance, by measuring its weight as a fraction of the total soil weight (gravimetric method). However, this measurement method usually is destructive since the soil sample is removed from the field to be analyzed in the laboratory. Moreover, it is a time-consuming and impractical way of measuring SWC in the field. Because of these limitations, a variety of indirect measurements (also called surrogate methods) have been developed.

An indirect method measures another variable that is affected by the amount of soil water, and then it relates the changes of this variable to the changes in SWC, through physically based or empirical relationships called calibration curves. For instance, the dielectric sensors exploit the changes in soil dielectric properties as function of SWC; the heat dissipation and heat flux sensors use the changes in the soil thermal properties; the neutron scattering technique is based on the loss of high-energy neutrons as they collide with other atoms, in particular hydrogen contained in the water molecule. Although the direct gravimetric method is the reference method for SWC measurement (and commonly used for indirect methods calibration), the majority of the commercial sensors are based on indirect methods. Specific descriptions are provided below.

a) Thermogravimetric measurement is a direct method, and it is the reference method for SWC measurement. It is based on the weight measurement of a wet sample before and after oven drying at 105 °C for 24 h. The difference in weight [the weight of liquid water (ml)] is expressed as fraction of the soil solid weight (ms), called gravimetric water content $(w = m_1 / m_s)$. This quantity can be expressed as volume fraction, by multiplying the gravimetric water content by the bulk density of the sample and dividing by the density of liquid water, $\theta = w\rho_b / \rho_1$, where ρ_b and ρ_1 are the soil bulk and water density, respectively. The value of bulk density should be obtained by volumetric and weight measurements of the same sample, which are then used to determine the gravimetric water content. Using bulk density values obtained from tables or from previous measurements should be avoided since SWC and bulk density are properties that vary in space and time. Both error (bias) and imprecision (larger variance) occur when volumetric water content is calculated using an assumed bulk density or one measured elsewhere or at another time. When the thermogravimetric measurement is performed for calibration of other SWC sensors, it is important to measure the sample bulk density for conversion into the volumetric form since many indirect methods (e.g., the dielectric sensors) provide volumetric measurements of SWC.

b) Dielectric measurement takes advantage of the differences in dielectric permittivity values between different soil phases (solid, liquid, and gas). Liquid water has a dielectric permittivity of ≈80 (depending on temperature, electrolyte solution, and frequency), air has a dielectric permittivity of ≈1, and the solid phase of 4 to 16. This contrast makes the dielectric permittivity of soil very sensitive to variation in SWC. The measurement of the bulk dielectric permittivity is then used to obtain the volumetric water content through calibration curves.

Although many different electronic devices and experimental techniques are available, all the dielectric sensors exploit the effect of liquid water dielectric permittivity on the bulk soil dielectric properties. Some sensors derive the dielectric permittivity by measuring the travel time of an electromagnetic wave traveling back and forth on the probe, such as time-domain reflectometry, or by measuring the capacitance of the bulk soil. Other sensors measure the dielectric properties of the reflected electromagnetic wave in the frequency domain, to obtain the dielectric properties of the bulk soil. Indeed, important families of sensors used for SWC are the frequency-domain reflectometry and the capacitance sensors, also referred as dielectric sensors or electromagnetic sensors. These sensors measure the dielectric permittivity of the media using capacitance/frequency-domain technology. Some devices are also equipped to measure SWC, soil temperature, and soil electrical conductivity (EC) within the same sensor.

The accurate use of these sensors requires a good understanding of several factors affecting the measurement, such as the geometric properties of the sensors, soil temperature, bulk soil EC, and the electronic features of the different sensors. Some authors analyzed different dielectric sensors to review their performance and compared them. Soil dielectric properties are also used as a basis for measurements of earth soil moisture using ground-penetrating radar (GPR) for measurement of SWC at larger scales.

c) Resistivity measurement is based on the principle that soil resistivity is affected by SWC. Usually, a current is transferred into the soil by electrodes, and the value of soil resistivity is then obtained by measuring the changes in voltage. The most common approach is to use four-probe resistance methods such as the typical Wenner array or other configurations that allow for insertion of multiple electrodes into the soil to obtain soil tomography. Other new technologies include the automatic resistivity profiling system, where the electrodes are the wheels of a machine pulled on the soil, which allows for rapid tomography of large areas. Additional resistivity methods are also available, such as the OhmMapper system, which uses electrodes that are dragged on the soil surface.

d) Neutron scattering technique also called neutron probe employs high-energy neutrons produced by a radiation source, which collide with soil atoms. Fast neutrons are emitted by a radioactive source. The neutrons lose their energy as they collide with other atoms, in particular hydrogen. Therefore, the neutrons are slowed down and counted. The instrument is equipped with a source of fast neutrons and a detector of slow neutrons. The number of hydrogen atoms in soils changes because of the change in SWC; therefore, the hydrogen content can be calibrated vs. the count of slow neutrons. According to a field-calibrated neutron moisture meter is the most accurate and precise indirect method for SWC measurement in the field. These sensors can be placed on the soil surface or as inserted tube for the measurement of the SWC profile.

e) Measurement of soil thermal properties is an indirect method that exploits changes in soil thermal properties due to variation of SWC. The two main techniques are heat dissipation and heat pulse. The heat dissipation technique uses a heat source (usually a heated needle) and temperature sensors (thermocouples or thermistors), immersed into a porous ceramic that equilibrates with the surrounding soil at a given water content. The needle is heated, and the rate of heat dissipation is measured by the temperature sensors. These changes are affected by the thermal conductivity, which depends on the ceramic water content. The thermal conductivity is then obtained through measuring the differential temperature before and after heating. In the heat flux method, the pulse of heat is applied at one location and its arrival at another location is determined by measuring the soil temperature at the other location. The time required for the pulse of heat to travel to the second location is a function of soil thermal conductivity, which is related to water content. The heat dissipation sensors are also used to estimate soil water potential, through calibration of the sensors at specific soil water potentials.

Although the techniques described above are the most common ones, other techniques are also developing such as acoustic wave methods, optical methods and gravity measurements.

Water Content Measurement at Different Spatial Scales

Determination of SWC can span many orders of magnitude ranging from 0.1 m², for common local measurements in the field employing in situ sensors, to 25 to 25,000 km², for satellite measurements. Since the measurement depth may vary from technique to technique, SWC determinations are usually expressed as volumetric measurements. But here, for simplicity, units of an area are used throughout the text. Intermediate scales can estimate SWC over areas of 10 to 100,000 m², using, for instance, GPR or electrical resistivity methods, to areas of 1 to 100 km², using aircraft remote sensing.

Local Scale

Measurement at the local scale ($\approx$0.01 m²) is performed with in situ measurement using sensors of different size and shape. As described above, among the most common devices for SWC measurement at this scale are the ones based on dielectric measurements. One of the main advantages of using in situ sensor is that these sensors can be connected to dataloggers and automatically retrieve SWC data in real time and provide detailed time series.

Figure depicts daily measurement of SWC at 10-cm depth, in a small catchment in northern Idaho, determined with a reflectometer probe. These sensors can provide very informative SWC time series at different depths, and they can be used to manage irrigation scheduling, compute the soil-water budget, corroborate soil–plant models, and other applications.

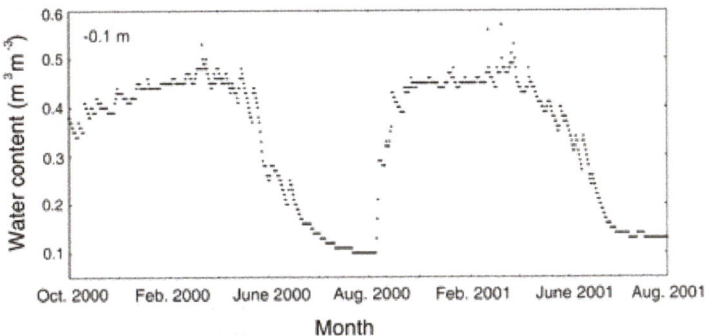

Figure: Example of local-scale soil water content measurement performed with time-domain reflectometer at a depth of 0.1 m from Oct. 2000 to Aug. 2001, in a vertical profile for a watershed in northern Idaho. Note the typical seasonal cycle of soil water content; 1 m = 3.2808 ft, 1 m³ = 35.3147 ft³.

Today, it is also possible to install distributed wireless sensor networks to obtain data in various locals across a field or a watershed. The current disadvantages are as follows:

1. Sensor calibration is often needed for use in different soil types,

2. Methodologies are not standardized, and

3. The SWC measurement is obtained only for very small soil areas (local measurement).

An important issue limiting the applicability of these sensors is the effect on the dielectric measurement of the dielectric losses occurring in saline and clay soils. For example, the main assumption behind the use of TDR is of negligible losses; therefore, assuming that only the real part of the dielectric permittivity determines the value of the TDR-measured apparent dielectric permittivity. This assumption is not valid for conductive soils (clay soils) or where high concentrations of electrolyte are present in the soil solution (saline soils) because, under these conditions, the contribution of the imaginary part is important. One of the main effects of dielectric losses on the TDR measurement is overestimation of SWC. More complex calibration equations were presented to address the issues of interferences from soil bulk EC and bound water related to soil temperature and clay content and type.

The spatial variability of SWC across a field is another important factor. One of the first researches analyzing the spatial and temporal variability of soil moisture was presented by Vachaud. In their work, they presented the concept of time stability. This concept describes that, for the same individual location, the SWC value maintain its rank in a cumulative probability function and its statistical parameters, for different sampling time. The rationale behind this concept is that, for similar vegetation conditions, the SWC depends on soil texture; therefore, this stability is based on the relationship between SWC and soil texture. Several improved estimation techniques have been developed based on the temporal stability concept, including the interpretation of measurements of SWC from microwave satellites. One of the largest studies on the spatial and temporal variability of SWC was the Southern Great Plains 1997 (SGP97) study. The statistical properties within fields (spatial variability) were related to the amount of SWC, with negatively skewed/nonnormal distribution under very wet conditions, normal in the midrange to positively skewed/nonnormal under dry conditions

Field Scale

Estimation of SWC at the field scale may range from a few meters (10 to 50 m²) up to several hectares (10,000 to 100,000 m²), and they are usually performed with geophysical methods such as GPR or electrical resistivity. GPR is a noninvasive technique that can detect subsurface structure and has been used to estimate SWC. Figure shows an increase in SWC after irrigation obtained using GPR ground wave and TDR measurements in a 60 × 60 m² area and at a depth of 0.1 m. Dabas presented a system for fast mapping of electrical resistivity by using a motorized vehicle to map large areas and then deriving the SWC from soil EC.

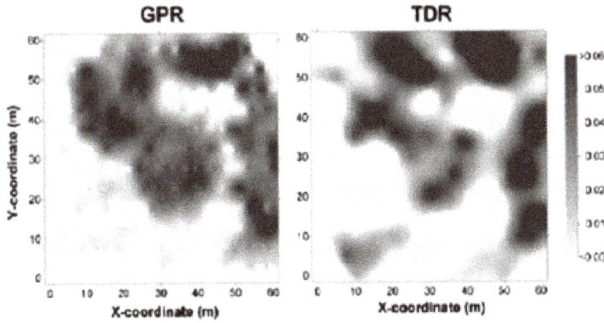

Figure: Example of field-scale maps, representing soil water content measured with ground-penetrating radar (GPR) and multiple time-domain reflectometer (TDR) in a 60 × 60 m² area; 1 m = 3.2808 ft.

Geophysical methods have the advantages of estimating SWC over areas that are much larger than the ones obtained from traditional local scale measurements and providing information for a few meters in depth. The disadvantages of geophysical methods is that they require regular field surveys to obtain information of SWC with time, whereas local and global scale techniques can provide automated acquisition, by either using automatic dataloggers (for in situ sensors) or periodic acquisition of satellite data. Moreover, the drawbacks of geophysical methods are commonly related to the difficulties of obtaining SWC from reflections measurements and the high cost of instrumentation such as GPR or resistivity methods.

Catchment Scale

The catchment scale is the most problematic scale since it is between the local and field scale and the regional and global scale. Currently, numerous local-scale sensors may be installed to obtain a network of local data; however, this approach is not cost-effective and geographic range is limited.

Geophysical methods (resistivity and GPR) can map larger sections of the catchment in a less labor-intensive manner. Aircraft remote sensing may be employed for large catchments, using microwave sensors mounted on the plane. The spatial resolution of airborne data typically ranges from 50 m to 1 km, while the SWC estimation can cover areas from 1 to 100 km². This intermediate spatial scale can be used to bridge the gap between local and field scale measurements and satellite observations. Specifically, airborne data can be used to assess the models developed from ground-based data and develop parameterizations that include the subpixel surface heterogeneity. Indeed, modeling techniques are very important for supporting data interpretation and for addressing the many unresolved issues that still limit applications of SWC measurement at this scale (such as down- and upscaling techniques and analysis of spatial heterogeneities). Determination of SWC from airplanes and helicopters using microwave remote sensing was presented by Macelloni, with scattering coefficients used to determine soil moisture. Santi presented a study where an airborne campaign with multifrequency microwave radiometers at L, C, and X bands was carried out on a flat alluvial area in northern Italy, in the Po Valley, with the aim of improving the estimation of airborne soil moisture and vegetation biomass maps, using both active and passive sensors. Ground measurements performed with TDR were used for testing and comparison of SWC data against the airborne data. Figure below shows an example of flight lines for the study of Santi, performed to measure brightness temperature from which estimation of SWC was obtained. Other studies on SWC determination from L-band and IR sensors from airplanes were presented by Merlin and Teuling.

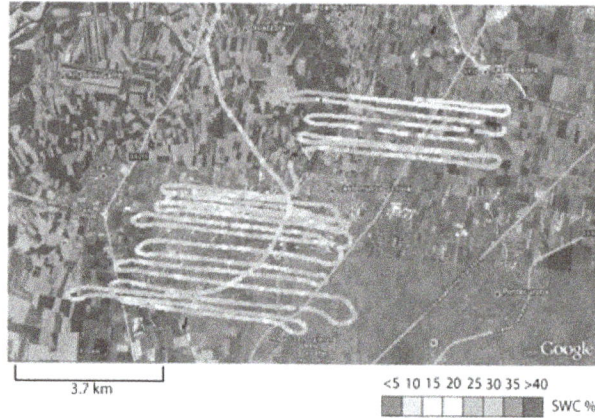

Above figure show example of geo-referenced flight lines performed by the aircraft to measure brightness temperature. The site selected for performing the experiments was a flat agricultural area close to Alessandria, northern Italy (central coordinates: lat. 45°N, long. 8.5°E), a flat alluvial plain measuring ≈300 km^2 (115.8 miles^2). Colors correspond to soil water content (SWC) values obtained from brightness temperature 1 km = 0.6214 mile.

Regional and Global Scale

Since the 1970s, various methods have been proposed for remote sensing of SWC, including visible and thermal space-borne data and microwave remote sensing. Microwave remote sensing using satellites has become the primary remote sensing technique for measurement of SWC at the regional and global scale.

Microwave instruments are well suited for soil moisture retrieval because of the strong relationship between dielectric permittivity and soil moisture. Moreover, microwave remote sensing is not significantly affected by cloud cover and is able to penetrate (to some extent) vegetation and soil while maintaining sensitivity to SWC. The sensors can be divided in two categories: passive sensors and active sensors. Passive sensors (radiometers) detect radiation emitted by the earth's surface, whereas active sensors (radars) transmit an electromagnetic wave to the target and measure the reflected or scattered energy back to the sensor.

Radiometers measure the brightness temperature, which is affected by soil, vegetation, snow cover, surface roughness, and atmosphere.

There are primarily two types of radar currently being used for SWC retrieval: synthetic aperture radar (SAR) and scatterometers. SAR is coherent radar, where high resolution images are created from the backscatter signals. Scatterometers are microwave radar sensors that measure the normalized radar cross section of the surface, scanned from an airplane or a satellite. They were primarily developed for measurement of near-surface winds over the ocean, based on the fact that wind determines small-scale changes of the sea surface, affecting the sea-surface roughness and, therefore, the backscattering properties. In addition to their original purpose, scatterometers are now also being used for polar ice studies, vegetation coverage, and SWC measurements. A variety of scatterometers have been launched on board satellites, such as the NASA scatterometer (NSCAT) and the sensor advanced scatterometer (ASCAT) on board of the ESA meteorological operational satellite (MetOp-A) launched in 2006.

Scatterometers have the advantage of being active during day and night time and to be unaffected by cloud coverage, providing around-the-clock coverage. Commonly, the scattering coefficients measured from the sensors are used as input for physical models to derive specific properties of the surface. However, physical models can often be difficult to parameterize, so simpler moisture retrieval methods have been presented, such as those based on change detection, where SWC is computed based on differences over long-term computation between the lowest and the highest value of backscatter. One of the disadvantages of the microwave sensors in general is the coarse spatial resolution, which is 50 and 25 km, depending on the sensors.

Flow in Saturated and Unsaturated Porous Media

Flows through porous media are described by Darcy's Law:

$$q = -\frac{K}{\mu}\nabla p$$

Where q is the average velocity, μ the viscosity and ∇p the pressure gradient of the fluid. It is valid in permanent regime. Unsaturated flows are usually considered as quasi-stationary phenomena, a succession of stationary states. Considering a one-dimensional flow and the continuity equation

$$\nabla \cdot q = 0$$

the pressure space-distribution is linear. This is validated in saturated flows, with the flow rate Q increases, as seen on figure.

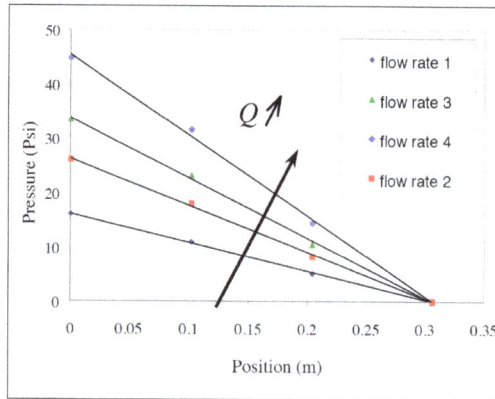

Figure: The pressure profile in saturated porous medium

The Unsaturated Porous Medium

The experiments were carried out on a rectangular mould used to measure unidirectional permeability. Additional pressure transducers were placed along the preform. In this case Darcy's Law can be assimilated to a purely resistive phenomenon and K to a hydraulic conductivity. In that case the preform was completely filled by the fluid. Similar measurements were made in unsaturated flows. As seen in figure below, the pressure distribution is not linear anymore but parabolic.

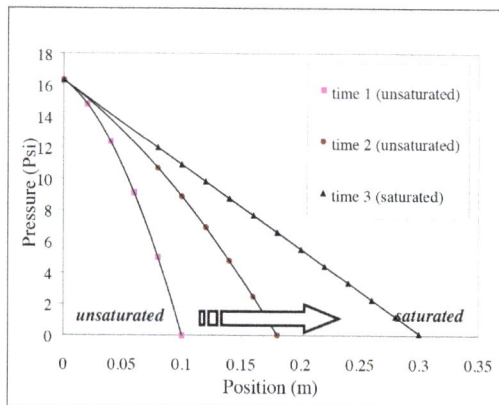

Figure: The pressure profile in unsaturated porous medium

This can be easily explained modifying the continuity equation and Darcy's equation as follow:

$$q = -\frac{K(S)}{\mu}\nabla p$$

$$\nabla \cdot q = -\frac{\partial Tr\varepsilon(S)}{\partial t}$$

where $Tr\varepsilon$ is the trace of the skeleton strain tensor that depends on the saturation. In fact, the permeability now depends on the saturation degree S since the skeleton's deformation depends on the amount of fluid in the pores (wet and dry preform). This saturation phenomenon is also taken into account in the continuity equation. In unsaturated flows the porosity variation term associated to fibre rearrangement within the fabric is neglectible regarding the first term involving the saturation degree variation. Hence in flow through a non-deformable porous medium, the problem simply is

$$q = -\frac{K(S)}{\mu}\nabla p$$

$$\nabla \cdot q = -\phi \frac{\partial S}{\partial t}$$

Solving this problem leads then to a parabolic pressure distribution as seen on figure below. Permeability dependence on saturation degree has been often discussed in the soil mechanics context, by Bear and is also observed in the LCM process by Spaid. Permeability values measured in unsaturated preform differ from those obtained on a saturated preform by a ratio that ranges between 0.4 for unidirectional reinforcement and 0.8 for fibre mats. This shows clearly the dependence of K on the saturation degree.

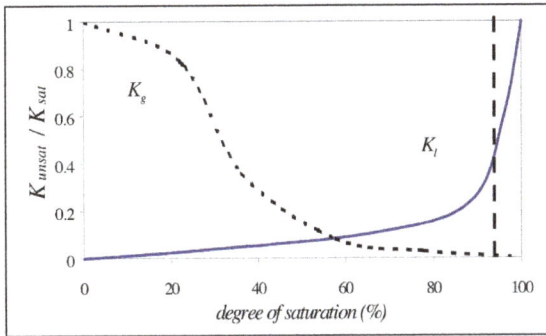

Figure: The saturation degree development

Hence K that was assimilated to a constant hydraulic conductivity should, in unsaturated flows, be considered as a variable conductivity depending on the saturation degree. The medium is defined by three components with the fibre element (f), the liquid element (l) and the gaseous element (g) representing the air bubbles. So, we can identify the liquid conductivity (K_l) and the gaseous conductivity (K_g) in figure above. The vertical line indicates that complete saturation of the porous medium is impossible. The flow through the porous medium is biphasic and hence defines biphasic saturation. However, we suppose that the capillarity effect is neglected and we consider that the flow is veritable monophasic behaviour with a saturation. The fact that $K_{sat} > K_{unsat}$ also explains qualitatively the convex shape of the pressure distribution in unsaturated flow in figure Since the permeability is smaller at the flow front, the local pressure gradient will be higher.

The Saturated Porous Medium

It is also interesting to study flows on a saturated preform. Darcy's Law considers only permanent regime. We concentrated our study on transient phases in saturated flows, where analysis has been developed in Bréard. As seen on figure below, following a flow rate change the pressure transient seems to be an exponential relaxation in unidirectional experiment.

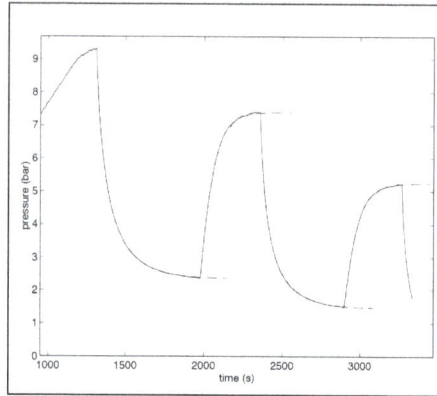

Figure: The double-scaled relaxation in saturated porous medium

We showed in fact that it is a double time-scaled dynamics consisting of the sum of two exponential functions.

$$p(t) = P_\infty - k_1 e^{-t/\tau_1} - k_2 e^{-t/\tau_2}$$

which represent flow in macro (τ1) and micropores (τ2) respectively, the faster transition corresponding to flow in macro-pores. It was also noted that these transient are not symmetrical which we think is the consequence of micro deformations of the porous media. Such experiments allow to quantify flow in macro/micropores with a macroscopic measurements. This form of exponential transients can now be obtained with the following equations, neglecting now the term of saturation degree variation (saturated porous medium) in the continuity equation and considering only effects due to a deformable porous medium:

$$q = -\frac{K}{\mu}\nabla p$$

$$\nabla \cdot q = -\frac{\partial Tr\varepsilon}{\partial t} = -C_1(\tau_1,\tau_2)\frac{\partial p}{\partial t} - C_2(\tau_1 \cdot \tau_2)\frac{\partial^2 p}{\partial t^2}$$

Where the constant C_1 and C_2 depend on the micro deformation of the structure. The total stress of the medium, due to compaction, is assumed to be constant during experiments. Moreover, there are non-linear effects in those behaviours that are integrated into constants C_1 and C_2 in saturated regime. However these disappear when stationary state is reached in a saturated medium.

The Intrinsic Permeability

Permeability in Darcy's Law should, in terms of modelling be considered as the product of the intrinsic geometrical permeability K_{geo} and a relative permeability $k_{rel}(S)$ depending on the saturation degree. It is, in fact, an intrinsic characteristic of the fibrous structure.

$$K = k_{rel}\left(S\right)K_{geo}$$

This highlights the point that, in fact, one should consider a triphasic medium: fibre, fluid, air. Indeed, the saturation degree is of primal interest for modelling, it can be a good prediction for default localisation since the less saturated the preform is, the more defaults are likely to happen. We can resume by the scheme on the figure below.

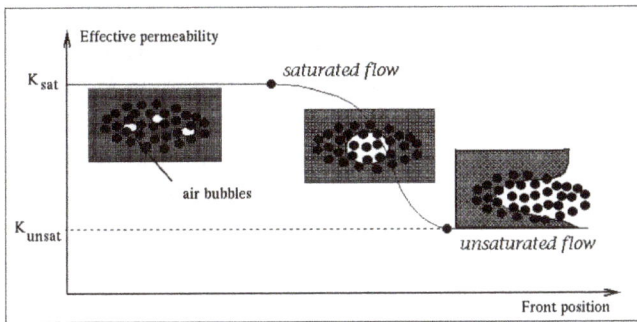

Figure: The flow behaviour through a dry preform

Soil Water Diffusivity Measurement

While dealing with most of the flow problems in hydrology and geoenvironmental engineering, the term soil water diffusivity becomes important. The unsaturated soil diffusivity is used for the moisture movement estimation in practice. One of the popular methods for diffusivity measurements is based on Mitchell's method which tests the soil suction value with time using thermocouple psychrometers.

Mitchell's Diffusivity Equation

Mitchell used a one dimensional flow to depict the moisture movement through the unsaturated soil, and defined a moisture source though the soil at a rate per unit volume as $f(x, t)$. The moisture into the soil is proposed by Mitchell:

$$\Delta Q = v_x \Delta y \Delta z \Delta t \big|_x - v_x \Delta y \Delta z \Delta t \big|_{x+\Delta x} + f(x,t) \Delta x \Delta y \Delta z \Delta t$$

Where Δx, Δy, Δz = dimensions of the specimen; v_x, v_y, v_z = the velocity of flow in three directions; and ΔQ = quantity of water based on the flow travel time Δt.

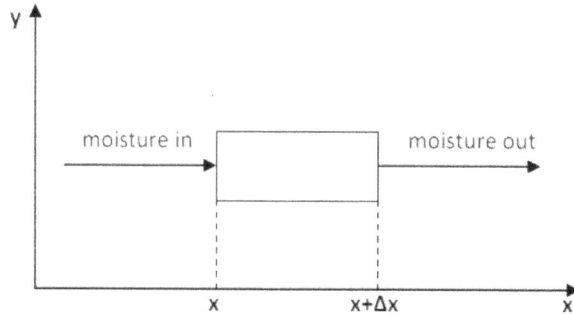

Figure: Moisture Flow

Based on the one dimensional flow in unsaturated soil, drying test was performed by Mitchell (1979). A cylindrical soil specimen with sealed surface and one open end from which the moisture evaporates to the atmosphere is shown in figure above. The moisture evaporates from the top surface where the atmospheric suction and the initial suction of the soil are known. The relationship between atmospheric suction ua and the initial soil suction ui at the soil surface $(x=L)$ is given by Mitchell:

$$\left(\frac{du}{dx} \right)_{x=L} = h_e (u_a - u_i)$$

Where h_e = evaporation constant. The boundary conditions of the drying test as shown in figure are as follows:

Initial suction : $u(x,0) = u_0$

Sealed boundary : $\dfrac{\partial u(0,t)}{\partial x} = 0$

Open boundary : $\dfrac{\partial u(L,t)}{\partial x} = h_e \left[u_a - u(L,t) \right]$

The one-dimensional solution for Equation $\left(\dfrac{du}{dx} \right)_{x=L} = h_e (u_a - u_i)$ with those boundary conditions is:

$$u(x,t) = u_a + \sum_0^\infty \frac{2(u_0 - u_a)\sin z_n}{z_n + \sin z_n \cos z_n} \exp\left(\frac{z_n^2 \alpha t}{L^2} \right) \cos\left(\frac{z_n x}{L} \right)$$

Where $u(x,t)$ = suction as a function of location and time; z_n = solution of $\cot z_n = (z_n/h_e L)$; h_e = evaporation coefficient; α = drying diffusion coefficient. The diffusion coefficient can be determined by measuring soil suction with time in a cylindrical soil specimen according to Equation $\left(u(x,t) = u_a + \sum_{0}^{\infty} \dfrac{2(u_0 - u_a)\sin z_n}{z_n + \sin z_n \cos z_n} \exp\left(\dfrac{z_n^2 \alpha t}{L^2} \right) \cos\left(\dfrac{z_n x}{L} \right) . \right)$

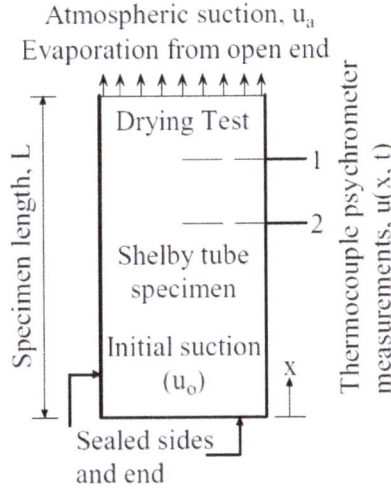

Figure: Boundary Conditions for Drying Test

Diffusivity Determined by SWCC and Permeability Coefficient

The diffusivity of unsaturated soils can also be expressed by the slope of the soil water characteristic curve (SWCC) and saturated permeability of soil:

$$\alpha = \frac{k_0 h_0}{c} \frac{\gamma_w}{\gamma_d}$$

where, k_o = saturated permeability of soil, h_o = a constant suction equal to 100 cm, γ_w = unit weight of water, γ_d = dry unit weight of soil, and c = slope of suction in logarithmic scale versus gravimetric water content.

In this equation, the diffusion coefficient can be derived from soil-water characteristic curve and coefficient of saturated permeability. The determination of SWCC and permeability coefficient will be presented, and the incremental measurements of diffusion coefficients will be discussed in the following sections.

Soil Water Characteristic Curve (SWCC)

The SWCC describes a nonlinear relation between the water content and the water potential in soil mass. The original models and analysis of the SWCC are from soil science and agriculture research. There are many conceptual models for SWCC equations, and the most frequently used model in geotechnical engineering practice was proposed by Fredlund and Xing.

The SWCC is mainly about the relation between the water content and the suction. In geotechnical engineering, the gravimetric water content, volumetric water content, and degree of saturation are used to plot SWCC. The volumetric water content is more preferable for geotechnical engineers, because the volumetric water content or degree of saturation is more representative for the analysis of the soil behavior associated with volume change.

There are two crucial transition points for SWCC: the air-entry value and the residual value. The air-entry value and residual value divide the SWCC into three zones: transition zone, boundary effect zone and residual zone. The air-entry value is the data point "where the air starts to enter the largest pores in the soil". The residual value is the point when a large suction is needed to remove the water in soil.

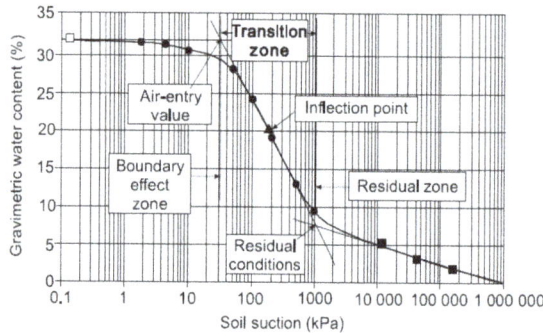

Figure: Desorption SWCC distinct zones

Empirical Equations for SWCCs

There are a large number of empirical equations for SWCC based on fitting laboratory data. In the early years, the gravimetric water content is used to describe SWCC. With the understanding of SWCC, the volumetric or degree of saturation has been applied to the empirical equations.

One of the earliest empirical equations to describe soil-water characteristics is given by Gardner:

$$\Theta_d = \frac{1}{1 + a_g \psi^{ng}}$$

where Θ_d = dimensionless water content $\left(\Theta = \dfrac{w}{w_s} \right)$, w = gravimetric water content of soil, w_s = saturated water content, a_g = fitting parameter (a function of air-entry value), n_g = fitting parameter.

Brooks and Corey proposed their best fit equation:

$$\Theta_n = \left(\frac{\psi_b}{\psi}\right)^{\lambda}$$

where Θ_n = normalized water content $\left(\Theta = \dfrac{w - w_r}{w_s - w_r}\right)$, w = gravimetric water content of soil, w_s = saturated water content, w_r = residual water content, ψ_b= air-entry value, λ= pore size distribution, and ψ = suction variable.

In Brutsaert's 1967 best fit curve, both fitting parameters and normalized water content have been used:

$$\Theta_n = \frac{1}{1 + \left(\psi \middle/ a_b\right)^{n_b}}$$

where a_b = fitting parameter (a function of air-entry value), n_b= fitting parameter.

Laliberte (1969) first proposed a triple-parameter empirical equation in which the fitting

parameters are related to pore-size distribution

$$\Theta_n = \frac{1}{2} erfc[a_1 - \frac{b_1}{c_1 + (\psi / \psi_b)}]$$

where a_1, b_1, c_1 = fitting parameters, ψ = suction variable, ψ_b= air-entry value, and erfc is Gauss error function.

Van Genuchten (1980) gave a closed-form expression for hydraulic conductivity, in which m is related to n through the equation $m = (1-1/n)$.

$$\Theta_n = \left[\frac{1}{1 + (p\psi)^n}\right]^m$$

where p, n, and m = three different soil parameters.

Boltzmann distribution has been applied to analyze SWCC by Mckee and Bumb (1984). Mckee and Bumb proposed another SWCC exponential function which is based on Fermi distribution:

$$\Theta_d = \cfrac{1}{1 + \exp\left[\cfrac{\psi - a_{m2}}{n_{m2}}\right]}$$

where a_{m2}, n_{m2} = fitting parameters, and ψ = suction variable.

The most frequently used empirical equation in geotechnical engineering practice is proposed by Fredlund and Xing (1994). This best fit curve covers the low suction range and high suction range with a correction factor:

$$\theta\left(\psi, a_f, n_f, m_f\right) = C(\psi)\cfrac{\theta_s}{\left(\ln\left[e + \left(\dfrac{\psi}{a_f}\right)^{n_f}\right]\right)^{m_f}}$$

where a_f, n_f, and m_f = curve fitting parameters, θ_s = saturated volumetric water content, θ = volumetric water content corresponding to a selected soil suction, e = a constant equal to 2.71828, and $C(\psi)$ is correction factor:

$$C(\psi) = 1 - \cfrac{\ln\left(1 + \dfrac{\psi}{\psi_r}\right)}{\ln\left[1 + \left(\dfrac{10^6}{\psi_r}\right)\right]}$$

where ψ_r = the suction at residual value.

There are difficulties for those equations to describe the SWCC (from 0 kPa to 1,000,000 kPa) with a continuous function. Pham and Fredlund (2005) proposed a piecewise linear equation to fit the SWCC data:

$$\begin{cases} w_1(\psi) = w_u - S_1 \log(\psi) & 1 \leq \psi \leq \psi_a \\[2mm] w_2(\psi) = w_a - S_2 \log\left(\dfrac{\psi}{\psi_a}\right) & \psi_a \leq \psi \leq \psi_r \\[2mm] w_3(\psi) = S_3 \log\left(\dfrac{10^6}{\psi_a}\right) & \psi_r \leq \psi \leq 10^6 \end{cases}$$

where S_1, S_2, S_3 = slope of SWCC at three zones, w_u = water content at 1 kPa, w_a = water content at air-entry value, w_1, w_2, w_3 = water content for the three zones.

Figure lists some of the SWCC empirical equations. Those equations are divided into three categories and the application of those equations depends on the suction range of the soil.

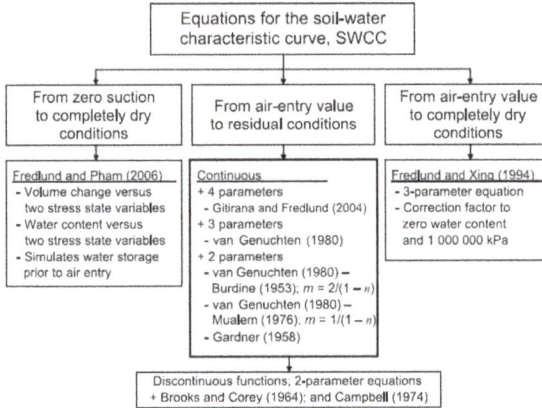

Figure: Categorization of SWCC

Permeability of Soil

The flow rate of water through a porous medium is regulated by the hydraulic conductivity (also called coefficient of permeability). For unsaturated soil, the water storage should be considered when coefficient of permeability is evaluated.

Unlike the coefficient of permeability in a saturated porous medium, the coefficient of permeability is a non-constant parameter related to the degree of saturation of the unsaturated soil (Brooks and Corey 1964). Proposed equations to evaluate the coefficient of permeability by using matric suction.

In this research study, the saturated permeability of the soil is used to analyze the diffusion coefficient. The tests of saturated permeability are based on Darcy's equation. The permeability test equipment amplifies the hydraulic gradient and quantity of water discharge from the soil to calculate the coefficient of permeability.

In practice, as shown in figure below the saturated permeability can also be evaluated by empirical correlations.

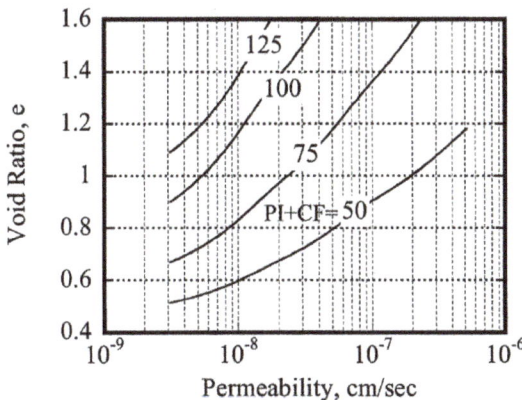

Figure: Empirical Correlations of Clay Permeability

Heat Flux through Soil

The amount of thermal energy that moves through an area of soil in a unit of time is the soil heat flux or heat flux density. The ability of a soil to conduct heat determines how fast its temperature changes during a day or between seasons. Soil temperature is a key factor affecting the rate of chemical and biological processes in the soil essential to plant growth. Soil heat flux is important in micrometeorology because it effectively couples energy transfer processes at the surface (surface energy balance) with energy transfer processes in the soil (soil thermal regime). This interaction between surface and subsurface energy transfer processes has led to detailed investigations of soil heat flux for a wide variety of agricultural systems.

Surface Energy Balance and Soil Heat Flux

In micrometeorology, measurement of soil heat flux is often considered within the context of the surface energy balance

$$R_n - G = LE + H$$

where R_n is the net radiation, G is the soil heat flux density at the soil surface, and LE and H are the latent and sensible heat flux densities, respectively. All terms in Eq. above have units of $J\ m^{-2}\ s^{-1}$ or $W\ m^{-2}$. Note that in Eq. above all fluxes away from the soil surface are defined as positive except for R_n. The left side of Eq. above, ($R_n - G$), represents the available energy while the terms on the right side (LE and H) are often referred to as the turbulent fluxes. Much of the energy that enters the soil during the day returns to the atmosphere at night through terrestrial longwave radiation. For this reason, G is often the smallest component of the daily surface energy balance and has, in some cases, been ignored; however, there are often significant transfers of energy into and out of a soil during both day- and nighttime hours and failure to include G in short-term (i.e., hourly) energy balance determinations can lead to sizeable errors.

Soil heat flux, as a component of the available energy, is a necessary input for many evaporation measurement and prediction techniques. Evaporation measured with the Bowen ratio energy balance approach (Bowen, 1926), for instance, is dependent on an accurate value for the available energy ($R_n - G$). The impact of errors in G on turbulent fluxes determined using the Bowen ratio method is discussed by Malek and de Silans. Several of the more common equations for predicting evaporation such as the Penman-Monteith and Priestley-Taylor, also require available energy as an input. The effect of omitting G on evaporation estimates will depend on local climate, soil properties, and cropping system. Although failure to include soil heat flux may introduce relatively small.

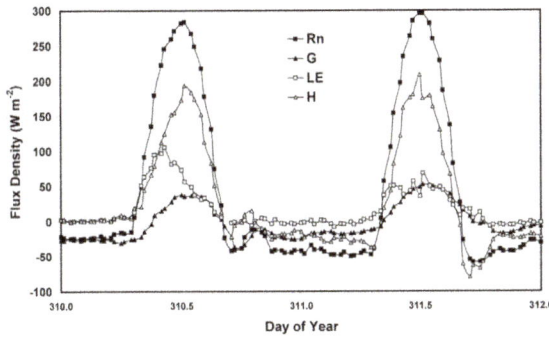

Figure: Diurnal patterns of energy balance terms for a no-till corn field in central Iowa. The corn residue layer was 0.05 m-thick. Soil heat flux density was measured at 0.05 m using the heat flux plate method and corrected for heat storage in the soil above the plate.

Soil Thermal Regime and Soil Heat Flux

In most instances, conduction is the principal mode of energy transport in soils, although radiation and convection in very shallow layers also may transfer energy. Heat flow in soil can be considered analogous to heat flow in a solid to which Fourier's Law is applied

$$G = -\lambda \partial T/\partial z$$

where λ is the thermal conductivity of the soil ($W\ m^{-1}\ K^{-1}$) and $\partial T/\partial z$ is the vertical temperature gradient ($K\ m^{-1}$) of the soil layer. Fourier's Law is defined for solid, homogeneous materials under steady-state conditions with thermal conductivities that are essentially constant over small temperature ranges. Fourier's Law is easily and directly used in many engineering applications, however, in a porous, three-phase medium like soil, use of Eq. above is considerably more difficult.

The thermal conductivity of soil varies by composition of the solid fraction (mineral type, particle size, and amount of organic matter), water content, and bulk density. These properties often vary between soils, spatially at the soil surface for the same soil, between layers within a soil, and over time. An eight-fold increase in the thermal conductivity of a silt loam soil as its gravimetric water content increased from oven dry to 0.34 kg kg^{-1}. A temperature gradient in the soil also induces water flow via evaporation and condensation, which include the concomitant transfer of latent energy. Solar radiation is the ultimate driving force behind soil heat transfer in most field settings. Thus, not only are soil thermal properties dynamic, but superimposed on these complex relationships are annual and diurnal patterns of solar radiation including irregularities in weather patterns.

Agricultural management practices including irrigation, drainage, and tillage have the potential to affect the thermal properties of soils and therefore soil thermal regime. In particular, the effect of tillage and crop residue management on soil heat flux has been the subject of several studies. Tillage loosens the surface soil, although some local compaction also may occur. Lower soil bulk density generally translates to lower λ, thus, lower G has been observed in tilled soil as compared with un-tilled or compacted soil. Crop residue has a low thermal conductivity and, whether lying on the soil surface or incorporated into the soil by tillage, may inhibit heat transfer into the soil. Residue layers also have a shortwave reflectivity that is higher than most soils and provide a barrier to vapor flow. Thus, soils with a large proportion of the surface covered by crop residue tend to have higher water contents, lower temperatures, and lower G. Such changes in soil thermal regime, of course, have implications for the surface energy balance and evaporation.

Measurement Techniques

Most soil heat flux density measurements have been completed using one of four methods (flux plate, calorimetric, gradient, or combination). One other method that has been developed but not been widely used is the block method. Measurement of soil heat flux density involves the measurement of soil temperature and one or more soil thermal properties (thermal conductivity or heat capacity), and possibly soil water content.

An area of current interest in soil heat flux measurement is the spatial variability of G under field conditions. Concern is heightened in energy balance studies where measurements of LE and H by eddy covariance or Bowen ratio techniques are typically representative of an area of 100s of m². By comparison, R_n (measured with a net radiometer) may be representative of an area of 10s of m² ; however, a single measurement of G is representative of a much smaller area, perhaps ~0.1 m². Several studies have shown that spatial variation of G under field conditions can be significant. This is especially true for sites with sparse canopies or uneven surfaces as both shading patterns and microtopography introduce large variation in soil temperature and soil water content. Variation in G (measured at 0.08 m depth) between adjacent locations with similar cover in a dune with an uneven surface and partial shrub cover to be greater than 200 W m^{-2}. In view of the evidence on spatial variability of G, measurement of soil heat flux density at multiple locations is necessary to obtain a spatially-representative value of soil heat flux during energy balance studies for agricultural surfaces, especially partial canopies.

Flux Plate Method

Most recent studies of soil heat flux density have used heat flux plates (also called heat flow meters or heat flow transducers). This trend is probably due to the comparative ease of the flux plate approach. The concept of a soil heat flux plate was adapted from efforts to measure heat transfer in walls of buildings and bulkheads of ships. Falckenberg is credited as the first to apply this approach specifically for measuring heat transfer in soil. Contributions to the advancements in theory and design of soil heat flux plates have been made.

Soil heat flux plates are small, rigid, wafer-shaped sensors that are placed horizontally into the soil. The plates make a direct measurement of heat flux density that is proportional to the heat flux density in the soil. Most designs employ an encapsulated thermopile, which produces an electromotive force (emf) in response to a temperature gradient across the plate created by vertical heat flow through the sensor body. Two designs of soil heat flux plates do not use a conventional thermopile as the sensing element. One uses a Peltier cooler and the other uses printed circuit technology. The signal from a plate that has been calibrated under conditions with known heat flow can then be used to determine the soil heat flux density at the depth of plate placement.

In spite of the simplicity and wide acceptance of the heat flux plate method, several areas of concern surround this approach: (i) heat flow convergence/divergence around the plate, (ii) water flow divergence (including water vapor), (iii) thermal contact between the plate and soil matrix, and (iv) accounting for heat storage in the layer(s) between the plate and the soil surface.

The presence of an impervious plate near the soil surface creates concern regarding convergence/divergence of heat flow, thereby biasing estimates of G obtained from the plate readings. Philip (1961) recommended several factors to consider in plate design to minimize perturbations in vertical heat flow based on theoretical analysis of heat flow near a small oblate spheroid of known thermal properties. An equation was derived to predict the ratio of heat flow in the soil to that through the plate:

$$G_m/G = \varepsilon/[1 + (\varepsilon - 1)H]$$

where G_m is the heat flux density through the plate, ε is the ratio of the plate thermal conductivity to the soil thermal conductivity (λ_m/λ), and H is an empirical factor related to plate shape:

$$H = 1 - (\beta b)$$

where β is a dimensionless geometric constant dependent on plate shape and b is the plate thickness divided by side length for a square plate or plate thickness divided by plate diameter for a circular plate. Mogensen (1970) tested Eq. $G_m/G = \varepsilon/[1 + (\varepsilon - 1)H]$ in laboratory experiments with a small circular plate and found that the heterogeneous composition of the plate made it difficult to verify Philip's theory. Watts et al. (1990) confirmed and refined Philip's recommendations to conclude that: (i) the plate thickness to width ratio should be small, (ii) dry and saturated sand are suitable calibration media for mineral soils, and (iii) the thermal conductivity of the plate should be greater than 0.5 W m^{-1} K^{-1}.

Figure 7–2 illustrates the effect of heat flux plate thickness and thermal conductivity on the ratio of the heat flux through the plate to the heat flux through the soil (G_m/G) as predicted by Eq. $G_m/G = \varepsilon/[1 + (\varepsilon - 1)H]$. The curves in figure. are for hypothetical circular plates that are 50 mm in diameter, 2, 4, or 6 mm-thick, and have thermal conductivity (λ_m) of 0.25 or 0.75 W $m^{-1} K^{-1}$. As the difference between the plate and soil increases, increasingly large errors (up to −50%) in measured G are predicted. Clearly, using a heat flux plate with λ similar to the expected soil λ or within a range common for mineral soils (~0.4–1.2 W m^{-1} K^{-1}) would reduce errors associated with heat flow convergence/divergence through the plate. The curves in figure also indicate that there is less error associated with thinner plates. For example, if a soil had a λ of 1.0 W m^{-1} K^{-1}, a 2 mm-diameter plate with a λ_m of 0.25 W $m^{-1} K^{-1}$ would have 22% less error in measured G than a 6 mm-diameter plate with the same λ_m. It should be noted that Philip's analysis assumes that the flux measurement is made across the entire plate area, while several current plate designs include a "guard" area surrounding a thermopile located in the center of the plate. Also, this analysis is only for heat conduction and does not include any provision for the effects of energy transfer due to liquid water movement near a plate.

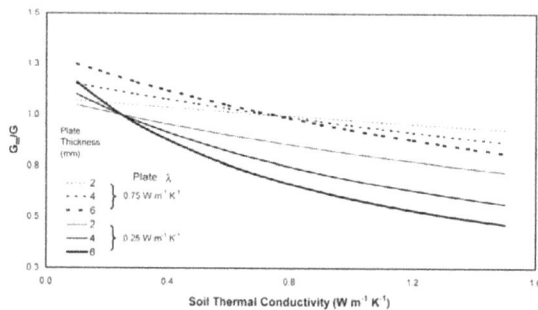

Figure: Estimated ratio of soil heat flux density measured with 50 mm-diameter round plates 2-, 4-, and 6-mm-thick with thermal conductivity of either 0.25 or 0.75 W $m^{-1} K^{-1}$ (G_m) to actual soil heat flux density in soil (G) with from 0.2 to 1.5 W $m^{-1} K^{-1}$ as predicted by Eq. [3].

The presence of an impervious object like a heat flux plate near the soil surface can affect water flow in the soil in several ways. During and after precipitation events, a heat flux plate can impede water movement downward in the soil and result in a higher water content in the soil immediately above the plate and lower water content immediately below. As evaporation dries the soil, the plate also can prevent upward water movement so the soil above the plate can become drier than the surrounding soil while the soil below the plate remains wetter. When soil immediately adjacent to the plate is at a water content that differs from the surrounding soil, it will also have different thermal properties. For a typical silt loam soil, a change in water content from 0.15 to 0.20 kg kg^{-1} can increase its λ from 0.96 to 1.3 W $m^{-1} K^{-1}$ and its volumetric heat capacity (C) from 1.75 to 1.96 MJ $m^{-3} K^{-1}$. Unrepresentative soil water content near a heat flux plate will affect the soil thermal properties and could result in sizable errors in measured G due to altered heat flow in the vicinity of the plate.

If a soil heat flux plate is positioned above the drying front, energy is consumed through the latent heat of vaporization (i.e., evaporative cooling) in the soil below the plate. This loss of energy may not be measured by the plate as the source of the energy may be from deeper in the soil, nor is it accounted for in the calorimetric heat storage correction, which is completed only for the soil layer above the plate Large errors in G (up to 100 W

m^{-2}) can occur when latent heat loss during periods of high evaporation from bare soil is ignored; however, such large errors are likely only when the plate is positioned near the surface (<0.02 m), evaporation is high, and the drying front is abrupt. If a plant canopy is present, there is a diffuse drying front, or the flux plate is at 0.05 m depth or greater, the error due to latent heat loss should not be significant.

Thermal contact between the plate and soil depends on soil texture, structure, water content, and care in installation of the plate. Fuchs and Hadas (1973) compared laboratory and field calibrations of soil heat flux plates and concluded that contact resistance was the largest source of error in the field measurements. To avoid errors associated with contact resistance, the calibration media should be selected so that the thermal contact is similar to that of the soil where the plate is to be installed. Alternatively, plate calibrations could be completed in situ by comparing G determined with flux plates with G determined by one of the other methods.

As the heat flux plate measures G only at the depth of placement, the calorimetric method has been used to account for the change in heat storage in the soil above the plate (Fuchs & Tanner, 1968). Failure to account for heat storage in the soil above the plate can result in large errors. Mayocchi and Bristow (1995) reported errors as large as 80 W m^{-2} in daytime, half-hour average G when the change in heat storage was ignored in a sugar cane (Saccharum officinarum L.) field in Australia. Massman (1992, 1993) concluded that the standard calorimetric correction may itself have errors of ±3 to 10% when assuming that the change in temperature of the soil layer above the plate is well-approximated by the average temperature at the midpoint. de Silans et al. (1997) present an analytical method for determining G at the surface from G measured by a plate and temperature at the soil surface and at the plate depth. This approach requires no knowledge of soil thermal properties but does require harmonic analysis of multiple days of steady, periodic temperature and heat flux waves.

Sensors

In general, plate design is dictated by the requirements that the plate be thin, watertight, have a λ_m comparable to the soil being monitored, and that the emf produced is high enough to be easily measured. To assure good thermal contact with the soil, exterior materials with a high thermal admittance are desirable. Thermal admittance (A_T) is defined as

$$A_T = (\lambda_m c_m)^{0.5}$$

Soil Heat Flux

where C_m is the heat capacity of the plate. To obtain a high A_T, several plate designs have used metal and/or anodized metal shields on their exteriors.

Due to the popularity of the flux plate technique, several sensors of varying dimensions, thermal properties, and sensitivities are now commercially available. Factors to consider when selecting a soil heat flux plate include the soil thermal and water regimes being monitored, and desired depth of placement. Plates with larger areas and lower λ_m positioned at 0.025 to 0.05 m depth may be suitable for G measurements with dry soils in arid environments as the soil λ and latent heat loss are likely to be relatively low. Smaller plates with higher λ_m positioned at 0.05 to 0.1 m may be preferable at a humid site where more frequent rainfall events keep the soil moist so the soil λ is higher and the drying front is above the plate depth.

Procedure

Figure illustrates one arrangement of sensors for application of the heat flux plate method. A shallow excavation to a depth below the desired depth of plate placement is made followed by a horizontal slit slightly smaller than the plate dimensions in a side wall. The plate should be carefully inserted into the slit so that the plate faces are parallel to the soil surface and there is good thermal contact with the soil on all sides. Depth of placement for soil heat flux plates is typically 0.025 to 0.1 m. At depths <0.025 m, there is concern that the soil may crack when dry thereby exposing the plate to solar radiation or creating poor thermal contact between plate and soil. Multiple plates in the same excavation or multiple excavations with single plates may be necessary to adequately represent the site being monitored. Distance between adjacent plates should be at least double the largest dimension (diameter or side length) of the plate face. Distribution of measurement sites should reflect any spatial heterogeneity induced by shading or microtopography. A length of sensor wire near the plate should be buried at the same depth as the plate to reduce the risk of heat conduction down the wire to/from the sensor.

Table: Specifications of some commercially available soil heat flux plates

Model[†]	Shape	Dimensions (L × W or diameter) mm	Thickness mm	Thermal conductivity W m^{-1} K^{-1}	Sensitivity µV/ W m^{-2}
CN3[‡]	rectangular	48 × 29	7	0.4	21
MF-81[§]	rectangular	110 × 12	4	0.4	26
HFP01SC[¶]	circular	80	5	0.8	50[¶]
GHT-1C[#]	square	52 × 52	5.7	0.26	900
HFT-1[††]	circular	38	3.9	1.0	24
610[‡‡]	circular	25	2.6	0.33	7.5
WS 31S[§§]	circular	110	5	0.2–0.3	100

† Names are necessary to report factually on available data; however, the USDA neither guarantees nor warrants the standard of the product, and the use of the name by USDA implies no approval of the product to the exclusion of others that also may be suitable.

‡ Carter-Scott Manufacturing Pty. Ltd., Brunswick, Victoria, Australia.

§ EKO Instruments Trading Co., Ltd., Tokyo, Japan.

¶ Hukseflux Thermal Sensors, Delft, the Netherlands.

International Thermal Instrument Co., Del Mar, CA.

†† Radiation and Energy Balance Systems, Seattle, WA. ‡‡ C.W. Thornthwaite Associates, Pittsgrove, NJ. §§ TNO Institute of Applied Physics, Delft, the Netherlands.

¶¶ This sensor can be used in a self-calibrating mode.

Soil temperature above the heat flux plates is often measured with thermocouples or thermistors to facilitate the calorimetric heat storage correction. In general, one to four measurement depths are used depending on the depth of the plate and uniformity of soil properties. Ideally, the temperature sensor profile should be located adjacent to each plate, however, depending on plate spacing and spatial variability of soil properties, the same temperature profile could be used for the heat storage correction of multiple plates in near proximity.

Soil Heat Flux Plate Calibration

Soil heat flux plates have been calibrated in situ against G determined by the calorimetric and gradient methods. An in situ calibration should reduce any errors associated with thermal contact resistance. Such calibrations also provide the best opportunity for obtaining accurate calibration data at varying soil moisture content; however, an in situ calibration involves a considerable investment in time and resources and measurement errors associated with the method used as the standard must be considered. One recent development in sensor technology is a heat flux plate that includes a heater for independent, in situ calibrations.

Several other techniques for soil flux plate calibration have been proposed including a radiation technique and several designs of steady-state laboratory apparatus. The laboratory techniques involve placing the plates in a porous medium (generally quartz sand) inside a well-insulated box where a known, 1-dimensional heat flux is maintained across the media and through the plates. Because establishment of a temperature gradient across moist sand creates a moisture gradient and latent heat transfer, accurate calibrations using this technique can be completed only under dry and saturated conditions.

Advantages and Disadvantages

The primary advantages of the flux plate method are that the plates are relatively inexpensive, reliable, and can be used continuously in the field for extended periods. For these reasons, the flux plate method has become the most popular technique for measuring G, especially in surface energy balance applications. Nonetheless, significant errors in measured G are likely unless certain precautions are taken and/or corrections made:

1. Plate thermal properties should be matched with the environmental conditions being monitored. As previously noted in figure, errors as large as 50%

are estimated from Eq. $G_m/G = \varepsilon/[1 + (\varepsilon - 1)H]$ when using plates with thermal conductivity grossly different than the surrounding soil. Plate thickness and face area along with near-surface soil water dynamics also should be considered when choosing the depth of placement.

2. Plates should be carefully calibrated and installed. Fuchs and Hadas (1973) measured a 27% difference in sensitivity for a flux plate calibrated both in the field and under controlled laboratory conditions. Most of this error was attributed to differences in thermal contact resistance with the soil.

3. Sufficient numbers of plates should be installed to obtain a spatiallyrepresentative estimate of G. Especially with partial canopies and uneven soil surfaces, spatial variation of G can be significant necessitating the careful placement of multiple sensors to obtain an accurate areal average.

4. Correction for heat storage above the plate must be made. Figure shows measured G at 0.05 m (average of three heat flux plates) at 0.5- h intervals with and without correction for heat storage. Although the daily sum of G for the corrected flux is only 3.3% greater than the uncorrected flux, the heat storage correction results in significantly greater G during the daytime and lower G at night. At 10:00 a.m. Central Standard Time (1000), the difference between the corrected and uncorrected G is 52 W m^{-2} (120–68 W m^{-2}). If the Bowen ratio energy balance method were being used to determined the turbulent fluxes, failure to correct for heat storage would lead to a 52 W m^{-2} (14%) overestimate of the available energy ($R_n - G = 377$ W m^{-2} instead of 325 W m^{-2}). This error would lead to a subsequent underestimate of $LE + H$ by the same amount.

5. Correction for latent heat loss may be necessary for shallow plates above an abrupt drying front. Buchan (1989) and Mayocchi and Bristow (1995) have shown that errors up to 100 W m^{-2} can occur if latent heat transfer is neglected under these conditions. This error is probably insignificant when plates are positioned below 0.05 m.

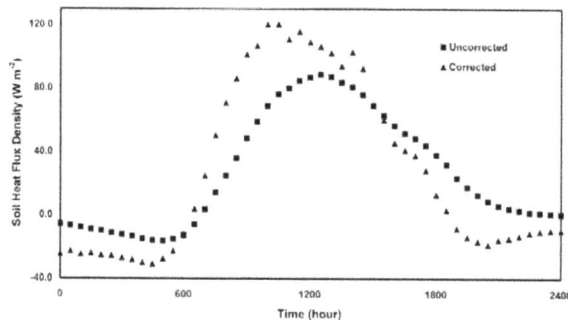

Figure: Heat flux density measured at 0.05 m in a silt loam soil in southwestern Minnesota with and without the correction for heat storage in the soil layer above the heat flux plate (T.J. Sauer and N.S. Eash, unpublished data).

Calorimetric Method

The calorimetric or temperature integral method is used to determine the average soil heat flux density from the change in heat storage in the soil over a given time interval:

$$G_{i-1} = \delta z_i C_i (\partial T_i / \partial t) + G_i$$

where G_{i-1} is the heat flux density at the top of a layer, δz_i is the layer thickness (m), C_i is the volumetric heat capacity for the layer ($J\ m^{-3}\ K^{-1}$), $\partial T_i/\partial t$ is the rate of change of the mean layer temperature (K), and G_i is the heat flux density at the bottom of the layer. If a layer n is found with $\partial T/\partial z = 0$, then by Eq. $G = -\lambda \partial T/\partial z$, $G_n = 0$ and the soil heat flux density for layer j is given by

$$G_{j-1} = \sum_{i=j}^{n} \delta z_i C_i (\partial T_i / \partial t)$$

When $j = 1$, Eq. above gives G at the soil surface. Alternatively, if G_i is known by some other method at a reference depth (z_r), Eq. $G_{i-1} = \delta z_i C_i (\partial T_i / \partial t) + G_i$ can be used to calculate G at all other depths. If G_r is calculated using the gradient method, this approach is referred to as the combination method.

Sensors

Soil temperature and volumetric heat capacity are the only data inputs necessary for completing the calculations involved with the calorimetric analysis.

Soil temperature must be accurately measured at several depths from near the surface to a depth where G is negligible for the desired measurement interval, generally at ~1 m for discerning diurnal patterns in G. Temperature sensors with different characteristics may be desired for the surface layers (i.e., more sensitive thermistors or smaller thermocouples), where temperature changes rapidly, than for deeper in the profile where the amplitude of soil temperature is dampened.

Techniques for determining soil heat capacity are discussed by Kluitenberg (2002). Methods for measuring soil heat capacity use calorimetry or heat pulse probes that can be used in laboratory or field settings. A technique to estimate volumetric heat capacity based on the volume fractions of the mineral, organic matter, and water components of a soil developed by de Vries (1963) also is widely used. To use the de Vries estimation technique, the soil porosity, organic matter content, and volumetric water content must be known. Thus, if bulk density and organic matter content are known for a soil, only measurement of the volumetric water content is necessary to estimate the volumetric heat capacity. Soil water content can be measured by several intrusive or non-intrusive techniques. Multiple sensors or water content samples collected from each site and depth may be used to improve spatial averaging.

Procedure

A relatively deep excavation will be required at the site(s) that will allow insertion of sensors into an exposed soil face. Selection of measurement locations should include consideration of the presence or absence of a plant canopy or residue layer, surface roughness, row spacing and orientation, and proximity to anomalous features (e.g., large rocks, tree roots, and compacted areas). Careful selection of depth intervals is essential to successful application of the method. Abrupt discontinuities in soil particle size, mineralogy, or bulk density within layers should be avoided to ensure uniform volumetric heat capacities and linear temperature gradients within each layer. Typically, larger changes in water content and temperature are observed near the surface, which result in progressively smaller contributions to G per depth increment with distance from the surface. Optimal sensor placement in a homogeneous soil should approximate a geometric progression with depth (e.g., 0.01-, 0.02-, 0.04-, 0.08-, 0.16-, 0.32-, 0.64-, 0.96-, and 1.28-m depths).

Measurement frequency and averaging interval of the temperature measurements will depend on the time constant of the temperature and heat capacity sensors, and objectives of the measurement protocol (e.g., hourly flux vs. daily sums vs. seasonal trends). Typically, data signals are averaged across 15- to 60- minute intervals. If heat capacity is being estimated, soil water content and bulk density values must be available for each layer between the temperature sensors. Identification of a layer with $G = 0$ or determination of G at a reference depth then allows calculation of G at all other depths and at the soil surface using Eq. $G_{i-1} = \delta z_i C_i (\partial T_i / \partial t) + G_i$.

Advantages and Disadvantages

The calorimetric method is constrained by the ability to accurately determine changes in volumetric heat capacity for each soil layer and each time interval. Under conditions with slowly changing soil water content, this may not present a significant challenge; however, for layered soils or when soil water content is changing rapidly (e.g., wetting front advance after a rainfall event), obtaining accurate heat capacity values is difficult. Cellier et al. (1996) provide a recent example of use of the calorimetric method to measure G including the development of an estimation technique using commonly-measured micrometeorological parameters.

Very precise soil temperature measurements (~0.02–0.1 K precision) are required for successful application of this method. This is especially important at deeper depths where the temperature gradient is applied to progressively thicker layers with large volumetric heat capacities. This also is important if data are required for shorter time intervals with smaller temperature changes (Hanks & Jacobs, 1971; Fuchs, 1986). Given the relatively stringent requirements for the temperature measurements, the large number of sensors, and their placement deep in the soil, the calorimetric method is most often used as a reference for comparison with other methods.

Gradient Method

The gradient method is a direct application of Eq. $G = -\lambda \partial T / \partial z$. A measured temperature gradient ($\partial T / \partial z$) is combined with an estimated or measured thermal conductivity (λ) to determine G. The simplicity of this approach is offset by difficulty in obtaining accurate λ measurements under field conditions. Like volumetric heat capacity, soil thermal conductivity changes significantly with soil water content, in this instance due to the large difference in λ between water and air (0.57 vs. 0.025 W m^{-1} K^{-1} at 283 K, respectively). As a result, λ values for soil layers in the field can change by a factor of two to five across commonly observed changes in water content.

Sensors

Bristow (2002) provides details concerning the measurement of soil thermal conductivity using both steady-state and transient methods. The single heat probe and heat pulse methods are best suited for in situ field measurements. Also included in the discussion is an approximation by de Vries (1963) for estimating soil thermal conductivity based on a phase mixing model and the weighted volume fraction of soil constituents. Data on soil water content will be required if the de Vries method is to be used for estimating λ. Thermocouples or thermistors are again suitable temperature sensors for the gradient method as they have sufficient accuracy and are easily logged for continuous measurements.

Procedure

Due to difficulties in accurately measuring λ near the soil surface, the gradient method is generally applied at a reference depth of at least 0.2 m. The calorimetric method can be applied to the surface layers above the reference layer to determine G at the surface. Coupling the calorimetric and gradient methods is referred to as the combination method. A shallower or deeper reference depth may be appropriate or possible for different soils depending on texture, moisture content, and the amplitude of the temperature wave. Placement of soil temperature, thermal conductivity, and/or soil water content sensors will depend on layering within the soil and overall objectives of the G measurements.

An excavation must be made to allow insertion of sensors at the desired depths and at locations appropriate for accurate representation of the measurement area. Multiple sensors on the same horizontal plane or multiple sensor locations will improve the spatial sampling. Sampling frequency and averaging intervals will depend on sensor characteristics, sampling objectives, and time scale of changes in soil moisture and temperature. The temperature gradient can be obtained by differentiating a smooth curve fit to the temperature data or by simply taking the average temperature difference across each layer over the averaging interval.

Advantages and Disadvantages

The gradient method is simple to employ, however, accurate measurement of λ in situ can be challenging. Recent improvements in thermal conductivity measurement techniques, especially for in situ measurements, now make application of the gradient method more attractive. Nonetheless, accurate estimates or measurements of λ near the soil surface are necessary if G is to be determined at the soil surface for use in energy balance or evaporation equations. Kimball et al. (1976) used four variations of the gradient method to measure G in a bare loam soil. They concluded that the method produced acceptable results with λ values estimated by de Vries' theory when a 0.2 m reference depth was used but not with a 0.05 m reference depth.

Combination Method

The coupling of the gradient and calorimetric methods is known as the combination method. In some instances, the coupling of the soil heat flux plate and calorimetric methods has been referred to as the combination method. Here, the former will be considered the combination method while the latter is considered a necessary heat storage correction for accurate application of the flux plate method.

The combination of gradient and calorimetric methods takes advantage of the limited measurements required for the gradient method and the accuracy of near-surface G measurements of the calorimetric method. The gradient method is used to determine G at some reference depth (typically 0.2 m) through application of Eq. $G = -\lambda \partial T / \partial z$ and the calorimetric method is applied to successive layers between the reference depth and the soil surface using Eq. $G_{i-1} = \delta z_i C_i (\partial T_i / \partial t) + G_i$. The combination method avoids the deep profile measurements of soil temperature and volumetric heat capacity of the calorimetric method and enables the determination of G at the soil surface based on a flux determined from the gradient method applied to one layer.

Sensors

Appropriate soil temperature, thermal conductivity, and heat capacity sensors/techniques to be used with the combination method or its null-alignment variant have already been described for the calorimetric and gradient methods.

Procedure

Only a shallow excavation is necessary as all required sensors will be at or less than approximately 0.25 m. Again, choice of location(s) and number of sensors at each depth will be dictated by the heterogeneity of soil properties and surface cover within the area to be represented by the measurements. The number and vertical spacing of temperature sensors above the 0.2 m reference will be determined by the soil properties although a geometric progression from the surface is still advisable with maximum vertical separation of < 0.05 m. The temperature gradient at the 0.2 m reference depth can be determined from sensors placed above and below (e.g., 0.15 and 0.25 m). Unless the heat capacity is directly measured, soil water content and bulk density need to be measured or estimated for each layer concurrent with the temperature measurements to determine the volumetric heat capacity for the calorimetric calculations.

Advantages and Disadvantages

The heat storage calculations are applied to relatively thin layers near the surface so 0.1 K resolution in soil temperature measurements are acceptable. As the necessary soil temperature measurements are straightforward, success with the combination approach will depend on how accurate the thermal conductivity and volumetric heat capacity of the soil layers can be measured or estimated. Pikul and Allmaras (1984) and de Vries and Philip (1986) compared the nullalignment method with the theory of Philip and de Vries (1957) and reported contrasting results. Pikul and Allmaras (1984) found poor agreement between the null-alignment and mechanistic approaches especially for shallow soil layers and dry soil conditions. De Vries and Philip (1986) concluded that the failure to account for latent heat loss in the upper soil layers resulted in serious underestimation of λ and systematic errors with the null alignment technique.

Estimation and Prediction of Soil Heat Flux Density

The importance of soil heat flux to surface energy balance and evaporation investigations has encouraged the development of estimation and prediction techniques for use when measured values of G are unavailable. These techniques rely on surrogate micrometeorological data, parameters obtained using remote sensing technology, or information on soil thermal properties, for input into estimation algorithms. One simple and popular technique involves the ratio of G to R_n. Table lists example values of G/R_n that were obtained for a variety of agricultural surfaces. As expected, G/R_n is relatively high (>~0.15) for bare soils and sparse canopies and low for fully-developed crop canopies and forests, which have greater attenuation of R_n within the canopy. While the G/R_n ratio has proven to be consistent in some instances and especially during daytime hours (Fuchs & Hadas, 1972), the ratio has been found to be sensitive to changing soil water content and canopy density (Idso et al, 1975; Clothier et al., 1986; Oliver et al., 1987). For these reasons, using G/R_n ratios to estimate G are primarily useful as a first approximation, with the understanding that such estimates may have large errors relating to variation in surface characteristics over time.

Table: Typical values for the ratio of soil heat flux density to net radiation (G/Rn) for various agricultural surfaces.

Surface/canopy	G/R_n	Location(s)	Comments	Reference
Loam soils	0.22–0.51	Arizona and Montana	G/R_n increased as soil dried	Idso et al., 1975
Loess soil	0.34	Israel	G/R_n unaffected by soil wetness	Fuchs & Hadas, 1972
Silty clay soil	0.14	Syria	daylight hours	Oliver et al., 1987
Alfalfa	0.10	Arizona	midday values, $G/R_n = 0.3$ for stubble	Clothier et al., 1986
Barley	0.12	Syria	daylight hours, < 50% ground cover	Oliver et al., 1987
Grass	0.04	California	$R_n > 0$, mown and well-watered	Meek et al., 1989
Maize	0.06	New York	daylight hours, 2.5 m-tall canopy	Brown & Covey, 1966
Mixed prairie	0.19	Saskatchewan	daylight hours, thick thatch layer	Ripley & Redmann, 1976
Orchard	0.04	West Virginia	$G/R_n = 0.07$ with coal dust amended soil	Sharratt & Glenn, 1988
Pasture	0.10	The Netherlands	daylight hours	DeBruin & Holtslag, 1982
Pine forest	0.04	Australia	daylight hours, 7.5 m-tall canopy	Denmead, 1969
Pine forest	0.01	United Kingdom	daylight hours, 9 m-tall canopy	Oliver et al., 1987
Sorghum	0.11	Texas	daylight hours	Szeicz et al., 1973
Sugar beet	0.03	Nebraska	daylight hours, full canopy	Brown, 1976
Wheat	0.07	Australia	daylight hours, 0.32 to 0.55 m-tall canopy	Denmead, 1969

In spite of the uncertainty surrounding estimates of G from the G/R_n ratio, it is still reasonable to expect that soil heat flux would be some small percentage of net radiation, a flux that is relatively easy to measure. Plant canopy properties that affect soil heat flux include the height of the canopy, the leaf area index (LAI), and the amount of vegetative cover. Several techniques have been developed to improve estimates of G from R_n or other data by incorporating additional attributes that characterize canopy properties. Anadranistakis et al. (1997) investigated the relationship between G/R_n and LAI for a barley (Hordeum vulgare L.) crop at various stages of development in Greece. An exponential relationship was found between LAI and G/R_n for daytime hours. The G/R_n ratio was 0.43 when the LAI was near 0 and approached a limit of 0.1 for large LAI. Clothier et al. (1986) found inclusion of crop height and a spectral vegetation index (ratio of near infrared to red reflectance, NIR/Red) both improved estimates of G from G/R_n. Use of the NDVI resulted in improved estimates of G from R_n for fields with bare soil and cotton (Gossypium hirsutum L.) and alfalfa (Medicago sativa L.) canopies at different stages of growth. Kustas et al. (1993) explored nonlinear relationships between G/R_n and vegetation indices (VIs). They concluded that a power function

$$G/R_n = aVI^b$$

where a and b are fitted constants, was more appropriate than the previously derived linear relationships between G/R_n ratios and VIs such as the NDVI and NIR/Red reflectance.

Another approach for predicting G uses soil temperature measurements in combination with different solutions to soil heat flow equations. Since soil temperature is easily measured and continuous data records for multiple depths are often available, these techniques offer another opportunity to estimate G when direct measurements are lacking. Many of these techniques do, however, involve various assumptions concerning soil thermal properties and may be limited to certain prescribed conditions (e.g., homogeneous soil, constant soil water content, no canopy, sunny days).

Horton and Wierenga (1983) developed a method to estimate G that uses a harmonic analysis of soil temperature at one depth, if soil thermal diffusivity ($\alpha = \lambda/C$) is known, or at two depths if α is unknown. A shallow soil temperature can be described:

$$T(z, t) = T_t \sum_{n=1}^{M} A_{on} \sin(n\omega t + \phi_{on})$$

where T_t is the temporal average soil temperature, M the number of harmonics (usually 1 to 3 harmonics are adequate), A_{on} and ϕ_{on} are the amplitude and phase angle, respectively, of the nth harmonic, and ω is the radial frequency equal to 2/P with P being the period of the fundamental cycle (24 h for diurnal cycles). Fitting Eq. above to the observed shallow soil temperature values provides T_t, A_{on}, and ϕ_{on}. Soil heat flux density can then be described as a function of time and depth:

$$G(z,t) = \sum_{n=1}^{M} \left\{ A_{on} C \sqrt{n\omega\alpha} \exp(-z\sqrt{n\omega/2\alpha}) \sin\left[n\omega t + \phi_{on} + (\pi/4) - z\sqrt{n\omega / 2\alpha} \right] \right\}$$

Equation above represents the soil heat flux density, positive downward in a homogeneous soil profile, with the temperature at the surface described by a Fourier series. To calculate G with Eq. above, one has to know the values for A_{on} and ϕ_{on} for the temperature at one depth, as well as α and C for the soil. Horton et al. (1983) describe how α can be determined from measurements of soil temperature at two depths. Soil heat flux density estimated with this harmonic method was in good agreement with G measured using the calorimetric method for a clay loam soil used a harmonic analysis, in this instance, with soil temperature normalized with respect to daily maximum and minimum soil surface temperature. This technique was used to predict G in soils having different tillage and residue cover conditions.

A finite-difference solution to the transient heat flow equation to estimate G from hourly soil temperature data at three depths. The finite-difference form of the transient heat flow equation for two layers (three nodes) can be written:

$$\lambda_2(\Delta \overline{T}_2)/\Delta z_2 - \lambda_1(\Delta \overline{T}_1)/\Delta z_1 = C(T_2^{j+1} - T_2^{j})\Delta z_3/\Delta t$$

where the superscripts (j and j+1) indicate time and the subscripts (1, 2, and 3) indicate node and layer number (increasing with depth). Soil volumetric heat capacity was measured once a day and a least-squares solution was used to estimate daily values for λ_1 and λ_2, which then enabled calculation of G for each layer using Eq. $G = -\lambda \partial T/\partial z$. The finite-difference method produced values of G that had smaller errors than the harmonic method when estimating G in silt loam soils in West Virginia and Alaska. Used a finite difference approach coupled with surface energy partitioning equations to calculate soil heat flux density. Their method required knowledge of soil thermal and hydraulic properties, and calculations were driven by meteorological inputs, e.g., wind speed, air temperature and humidity, and radiation.

Soil Temperature Variations

Soil temperature fluctuates annually and daily affected mainly by variations in air temperature and solar radiation. The annual variation of daily average soil temperature at different depths can be estimated using a sinusoidal function.

There are two types of Soil Temperature:

Daily and Seasonal Variation of Soil Temperature

a. There variations occur at the surface of the soil.

b. At 5 cm depth the change exceeds 10° C At 20 cm the change is less and at 80 cm diurnal changes are practically nil.

c. On cooler days the changes are smaller due to increased best capacity as the soils become wetter on these days.

d. On a clear sunny day a bare soil surface is hotter than the air temperature.

e. The time of the peak temperature of the soil reaches earlier than the air temperature due to the lag of the air temperature.

f. At around 20 cm in the soil the temperature in the ground reaches peak after the surface reaches its maximum due to more tune the heat takes to penetrate the soil. The rate of penetration of heat wave within the soil takes around 3 hours to reach 10 cm depth.

g. The cooling period of the daily cycle of the soil surface temperature is almost double than the warning period.

h. Undesirable daily temperature variations can be minimized by scheduling irrigation.

Seasonal variations of Soil Temperature

a. Seasonal variations occur much deeper into the soil.

b. When the plant canopy is fully developed the seasonal variations are smaller.

c. In winter, the depth to which the soil freezes depends on the duration and severances of the winter.

d. In summer the soil temperature variations are much more than winter in trophies and sub trophies.

Model Description

The annual variation of daily average soil temperature at different depths is described with the following sinusoidal function:

$$T(z,t) = T_a + A_0 e^{-z/d} \sin\left[\frac{2\pi(t - t_0)}{365} - \frac{z}{d} - \frac{\pi}{2} \right]$$

where $T(z,t)$ is the soil temperature at time $t(d)$ and depth $z(m)$, T_a is the average soil temperature (°C), A_0 is the annual amplitude of the surface soil temperature (°C), d is the damping depth (m) of annual fluctuation and t_0 is the time lag (days) from an arbitrary starting date (taken as January 1 in this software) to the occurrence of the minimum temperature in a year. The damping depth is given by $d = (2D^h/\omega)^{1/2}$, where D_h is the thermal diffusivity and $\omega = 2\pi/365$ d^{-1}.

Assumptions and Simplifications

The sinusoidal temperature model was derived by solving the following partial differential equation:

$$\frac{\partial T(z,t)}{\partial t} = D_h \frac{\partial^2 T(z,t)}{\partial z^2}$$

where $T(z, t)$ is the soil temperature at time t and depth z and Dh is the thermal diffusivity.

The following assumptions are employed in the derivation of the temperature model:

1. A sinusoidal temperature variation at the soil surface $z = 0$. That is

$$T(0,t) = T_a + A_0 \sin\left[\frac{2\pi(t - t_0)}{365}\right]$$

where T_a is the average soil temperature, A_0 is the amplitude of the annual temperature function, t_0 a time lag from an arbitrary starting date (selected as January 1 in this software) to the occurrence of the minimum temperature in a year.

2. At infinite depth, the soil temperature is constant and is equal to the average soil temperature.

3. The thermal diffusivity is constant throughout the soil profile and throughout the year.

References

- Zone-of-saturation-1199: corrosionpedia.com, Retrieved 10 June 2018

- Modelling-Flow-and-Transport-in-Unsaturated-Zone-298409455: researchgate.net, Retrieved 19 July 2018

- Moisture-content, Soils-Protocol, Laboratory: eng.ucmerced.edu, Retrieved 30 June 2018

- Soil-moisture-content-in-the-field, learning-centre, soil-moisture-monitoring, soil-plants-climate: mea.com.au, Retrieved 17 April 2018

- Soil-Temperature, software: soilphysics.okstate.edu, Retrieved 19 May 2018

4

Evaporation through Soil

The rate of evaporation from the soil surface is influenced by various soil characteristics, environmental interactions and tillage. This chapter discusses in detail the process of evaporation through the soil, and includes the topics relevant to this study such as soil water redistribution during evaporation, vapor flow through soil and soil transpiration.

Soil Evaporation

Soil evaporation not only determines partitioning of available energy between sensible and latent heat flux for bare soil surfaces but can also significantly influence energy flux partitioning of partially vegetated surfaces. This latter effect occurs via the impact of soil evaporation on the resulting surface soil moisture and temperature. These, in turn, strongly influence the microclimate in partially vegetated canopies, indirectly affecting plant transpiration. Over a growing season, soil evaporation can be a significant fraction of total water loss for agricultural crops. On a seasonal basis in semiarid and arid regions, soil evaporation can significantly alter the relative fraction of runoff to rainfall, which in turn has a major impact on the available water for plants. In deserts, in spite of its small magnitude, soil evaporation can introduce significant errors in meteorological forecasting if neglected.

The measurement of soil evaporation at field scale is typically obtained using standard micrometeorological techniques, namely Bowen ratio and eddy covariance methods. Traditionally, due to fetch and measurement requirements, under partial canopy cover conditions, these techniques are not able to partition the total evapotranspiration into its soil evaporation and plant transpiration components. A novel procedure for partitioning evapotranspiration through utilizing the measured high-frequency time series of carbon dioxide and water vapor concentrations has been developed and tested. This approach relies upon the simple assumption that contributions to the time series of carbon dioxide and water vapor concentrations are derived from stomatal processes (i.e., photosynthesis and transpiration), and nonstomatal processes (i.e., respiration and direct evaporation) separately conform to flux-variance similarity. Vegetation water-use efficiency is the only parameter needed to perform the partitioning.

Soil evaporation in partial canopy cover conditions varies spatially depending primarily on soil water distribution, canopy shading, and under-canopy wind patterns. These effects are magnified in row crops and under various irrigation techniques (e.g., drip irrigation). Soil evaporation can be measured using microlysimeters, chambers, time domain reflectometers (TDRs), a combination of microlysimetry and TDR, micro-Bowen ratio systems, or heat pulse probes. Given the high spatial variability in the driving forces under partial canopy cover conditions, these point-based measurements are difficult to extrapolate to the field scale. Therefore, models have been developed to estimate the contribution of soil evaporation to the total evapotranspiration process.

Measurement methods are described, and models of varying degrees of complexity are reviewed, focusing primarily on relatively simple analytical models, some of which provide daily estimates and can be implemented operationally.

Measurement Methods

Microlysimeters

Microlysimeters have been widely used to measure evaporation from the soil surface of irrigated crops. Typically, an undisturbed soil sample (a representative vertical section of the soil profile) is inserted into a small cylinder open at the top. The microlysimeter is inserted back into the soil with its upper edge level with the soil surface and weighed either periodically or continuously. Changes in weight reflect an evaporative flux. To eliminate vertical heat conduction through the microlysimeter cylinder and minimize horizontal heat flux in the deeper layers of the sample, polyvinyl chloride (PVC) has been found to be the most suitable material. The microlysimeter's dimensions are typically a diameter of ~8 cm and a depth of 7–10 cm. Theoretically, the microlysimeters provide absolute reference for soil evaporation, as long as their soil and the heat balance are similar to the surrounding area.

Chambers are used to directly measure the flux of gases between the soil surface and the atmosphere by enclosing a volume and measuring all flux into and out of the volume.

With infrared gas analyzers (IRGAs) becoming increasingly common, they are widely considered to be the method of choice today for chamber-based soil respiration and evaporation measurements. Chambers can be used in either of two modes to calculate fluxes:

1. In steadystate mode, the flux is calculated from the concentration difference between the air flowing at a known rate through the chamber inlet and outlet after the chamber headspace air has come to equilibrium concentration of carbon dioxide;

2. In the non-steady-state mode, the flux is calculated from the rate of increasing concentration in the chamber headspace of known volume shortly after the chamber is put over the soil.

In both modes, air is circulated between a small chamber that is placed on the soil and an IRGA. Typically, a soil chamber of 1 L volume is placed on a PVC collar of about 80 cm² area. This collar is inserted about 2 cm into the soil and secured to prevent movement when the chamber is placed on it. When the chamber is placed on the collar, circulation of air between the chamber and the external IRGA is induced by a pump, and the water vapor concentration is measured.

Soil water balance Soil evaporation (E) can be extracted from the water balance equation, provided that all other components are known:

$$E = I + P - R + F - \Delta S$$

where I is irrigation, P is precipitation, R is runon or runoff, F is deep soil water flux (percolation), and ΔS is change in soil water storage. For an experimental field site, irrigation and precipitation can be easily monitored, and runoff and runon may be controlled to near-zero amounts by diking. Deep soil water flux errors can be reliably estimated in several ways, with the most important being the monitoring or measuring of the soil water content well below the root zone. The change in soil water storage can be determined fairly accurately with profile measurements of soil water content over multiple depths at the beginning and end of a defined time period.

There are many soil water content sensors, all of which work by measuring a surrogate property that is empirically or theoretically related to the soil water content. TDR methods as compared to bore hole capacitance methods. Of the three optimal methods in the study, TDR is the only methodology capable of providing automated continuous measurements. However, other continuous measurement sensors such as frequency-domain reflectometers (FDRs), and time domain transmission (TDT) sensors are also emerging as options for long-term installations, given appropriate calibration.

Micro-Bowen ratio systems The Bowen ratio energy balance (BREB) approach is one of the simplest and most practical methods of estimating water vapor flux and has thus been used extensively under a wide range of conditions providing robust estimates. Use of the BREB concept enables solving the energy balance equation by measuring simple gradients of air temperature and vapor pressure in the near-surface layer above the evaporating surface. The BREB equation is:

$$\lambda E = \frac{Rn - G}{1 + B_0}$$

in which λE is the latent heat flux, R_n is the net radiation, G is the soil heat flux, and B_0 the Bowen ratio, which is found from measurements of temperature and vapor pressure at two heights within the constant flux layer. Assuming equal transfer coefficients for heat and vapor, the Bowen ratio is defined as:

$$B_0 = \frac{H}{\lambda E} = \gamma \frac{\partial T}{\partial e}$$

where H is the sensible heat flux, γ is the psychrometric constant, T is the air temperature, and e is the water vapor pressure.

Application of the Bowen ratio concept to measure bare soil surface evaporation was suggested by measuring temperature and vapor pressure close to the soil surface (e.g., at 1 and 6 cm). Compared to microlysimeters, the micro-Bowen ratio (MBR) system yielded good results over bare soil. The potential of the MBR approach was demonstrated by successfully measuring soil evaporation within a maize field. To date, test of the MBR is very limited, and further research is required to examine the performance of the technique under various environmental and agronomic conditions.

Heat Pulse Probes

A novel approach for measuring soil evaporation has been proposed, based on the soil sensible heat balance. In this approach, a sensible heat balance is used to determine the amount of latent heat involved in the vaporization of soil water following Gardner and Hanks:

$$LE = (H_0 - H_1) - \Delta S$$

where H_0 and H_1 are soil sensible heat fluxes at depths 0 and 1, respectively; ΔS is the change in soil sensible heat storage between depths 0 and 1; L is the latent heat of vaporization; and E is evaporation.

Typically, three-needle sensors like those described by Ren et al. are used, which are spaced 6 mm apart, in parallel. Temperature is measured by all three needles, and the central needle also contains a resistance heater for producing a slight pulse of heat required for the heat pulse method. At a given time interval (2–4 hours), a heat pulse is executed, and the corresponding rise in temperature at the outer sensor needles is recorded. Having measurements of soil temperature, thermal conductivity, and volumetric heat, evaporation is estimated from Equation ($LE = (H_0 - H_1) - \Delta S$).

Numerical Models Numerous mechanistic/numerical models of heat and mass flows exist and are primarily based on the theory of Philip and de Vries. However, they continue to be refined through improved parameterization of the moisture and heat transport through the soil profile. Some of these mechanistic models have been used to explore the utility of bulk transfer approaches used in weather forecasting models and in soil–vegetation–atmosphere models, computing field to regional-scale fluxes. These bulk transport approaches are commonly called the "alpha" and "beta" methods defined by the following expressions

$$LE_s = \frac{\rho C_p}{\gamma}\left(\frac{ae_*(T_s)-e_A}{R_A}\right)$$

and

$$LE_s = \frac{\rho C_P}{\gamma}\beta\left(\frac{e_*(T_s)-e_A}{R_A}\right)$$

where ρ is the air density ($\sim$1kg/m³), Cp the heat capacity of air ($\sim$1000J/kg/K), γ the pyschrometric constant ($\sim$65Pa), $e_*(T_s)$ the saturated vapor pressure (Pa) at soil temperature T_S (K), e_A the vapor pressure (Pa) at some reference level in the atmosphere, and R_A is the resistance (s/m) to vapor transport from the surface usually defi ned from surface layer similarity theory. For a, several different formulations exist with one of the first relating a to the thermodynamic relationship for relative humidity in the soil pore space, h_R

$$h_R = \exp\left(\frac{\psi g}{R_V T_S}\right)$$

where ψ is the soil matric potential (m), g the acceleration of gravity (9.8m/s2), and R_V the gas constant for water vapor (461.5J/kg/K). From Eq. 6, β can be defi ned as a ratio of aeodynamic and soil resistance to vapor transport from the soil layer to the surface, R_S, namely

$$\beta = \frac{R_A}{R_A + R_S}$$

Both modeling and observational results indicate that more reliable results are obtained using the beta method. Evaporation rates for a wider range of conditions,

$$LE_s = \frac{\rho C_P}{\gamma}\beta\left(\frac{ae_*(T_s)-e_A}{R_A}\right)$$

Unfortunately, there is no consensus concerning the depth in the soil profile to consider in defining the a and β terms. In particular, studies evaluating the soil resistance term R_S use a range of soil moisture depths: 0–1/2 cm, 0–1 cm, 0–2 cm, and 0–5 cm. Using field data and numerical simulations with a mechanistic model, they find that the 0–5 cm depth, which can be provided by the L-band microwave frequency, appears to be the most adequate frequency for evaluating soil evaporation.

Besides the depth of the soil layer to consider, the equations relating soil moisture to R_S have ranged from linear to exponential. Furthermore, observations and numerical models have shown that varies significantly throughout the day and that its magnitude is also affected by climatic conditions.

These are not the only complicating factors that make the use of such a bulk resistance approach somewhat tenuous. The soil water flux in the 0–9 cm depth to be very dynamic with fluxes at all depths continually changing in magnitude and sometimes direction over the course of a day. A simple energy balance model in which the soil moisture available for evaporation is defined using the depth of the evaporating/drying front in the soil. This approach removes the ambiguity of defining the thickness of the soil layer and resulting moisture available for evaporation. However, the depth of the evaporating surface is not generally known a priori, nor can it be measured in field conditions; hence, this approach at present is limited to exploring the effects of evaporating front on R_S type formulations.

Analytical Models

To reduce the effect of temporal varying, R_S, Chanzy and Bruckler developed a simple analytically based daily LE_S (E_D) model using simulations from their mechanistic model for different soil texture, moisture, and climatic conditions as quantified by potential evaporation (E_{PD}), as given by Penman. The analytical daily model requires midday 0–5 cm soil moisture θ, daily potential evaporation, and daily average wind speed (U_D). The simple model has the following form

Table: Bulk soil resistance formulations, R_S, from previous studies

R_S formula (s/m)	Value of coefficients	Soil type	Depth (cm)
$R_S = a\left(\dfrac{\theta_s}{\theta}\right)^n + b$	$a = 3.5$ $b = 33.5$ $n = 2.3$	Loam[a]	0–1/2
$R_S = a(\theta_s - \theta) + b$	$a = 4,140$ $b = -805$	Loam[b]	0–1/2
$R_s = R_{SMIN} \exp[a(\theta_{MIN} - \theta)]$	$R_{SMIN} = 10$ s/m $\theta_{MIN} = 15\%$ $a = 0.3563$	Fine sandy loam	0–1
$R_S = \dfrac{a(\theta_s - \theta)^n}{2.3 \times 10^{-4}(T_s / 273.16)^{1.75}}$	$n = 10$ $\theta_s = 0.49$ $a = 8.32 \times 105$ $n = 16.16$ $\theta_s = 0.392$	Sand	0–2
$R_s = \alpha\theta + b$	$a = -73,420 - -51,650$ $b = 1,940 - 3,900$ $a = 4.3$	Sand	0–2
		Sand	0–2 0–5
$R_S = \exp\left[b - a\left(\dfrac{\theta}{\theta_s}\right)\right]$	$b = 8.2$ $a = 5.9$ $b = 8.5$	Gravelly sandy loam	

$$\frac{E_D}{E_{PD}} = \left[\frac{\exp(A\theta + B)}{1 + \exp(A\theta + B)}\right] C + (1 - C)$$

Where,

$$A = a + 5\max(3 - E_{PD}, 0)$$
$$B = b - 5(-0.025b - 0.05)\max(3 - E_{PD}, 0) + \alpha(U_D - 3)$$
$$C = 0.90 - 0.05c(U_D - 3)$$

where the coefficients a, b, and c depend on soil texture and were derived from their detailed mechanistic model simulations for loam, silty clay loam, and clay soils. In figure, a plot of Eq. above is given for two soil types, loam and silty clay loam under two climatic conditions, namely, a relatively low evaporative demand condition with U_D = 1 m/s and E_{PD} = 2 mm/d and high demand U_D = 5 m/s and E_{PD} = 10 mm/d. Notice the transition from $E_D/E_{PD} \sim 1$ to $E_D/E_{PD} < 1$ as a function of θ varies not only with the soil texture, but also with the evaporative demand. The simplicity of such a scheme outlined in Eq. above needs further testing for different soil textures and under a wider range of climatic conditions.

Table: Values of desorptivity, D_E, evaluated from various experimental sites

Desorptivity D_E (mm/d$^{1/2}$)	Soil type
4.96–4.30	Sand
5.08	Clay loam
4.04	Loam
3.5	Clay
~4 to ~8[a]	Loamb
5.8	Clay loam
4.95[c]	Silty clay loamd
2.11	Gravelly sandy loam

[a]The magnitude of DE was found to have a seasonal dependency.

[b]Soil type was determined from texture-dependent soil hydraulic conductivity and matric potential equations of Clapp and Hornberger evaluated by Camillo and Gurney.

[c]This value was evaluated for a vegetated surface.

[d]Soil type was determined from texture-dependent soil hydraulic conductivity and matric potential equations of Clapp and Hornberger evaluated by Sellers.

The ratio E_D/E_{PD} as a function of θ illustrated in figure also depicts the effect of the two "drying stages" typically used to describe soil evaporation. The "first stage" (S_1) of drying is under the condition where water is available in the near-surface soil to meet atmospheric demand, i.e., $E_D/E_{PD} \sim 1$. In the "second stage" (S_2) of drying, the water availability or θ

falls below a certain threshold where the soil evaporation is no longer controlled by the evaporative demand, namely, $E_D/E_{PD}<1$. Under S_2, several studies find that a simple formulation can be derived by assuming that the time change in q is governed by desorption, namely, as isothermal diffusion with negligible gravity effects from a semi-infinite uniform medium. This leads to the rate of evaporation for S_2 being approximated by

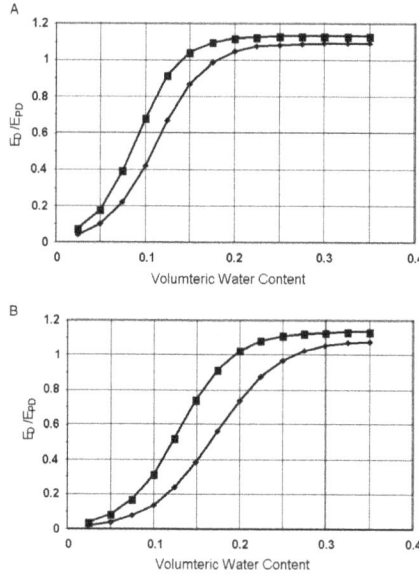

Figure: A plot of E_D/E_{PD}, estimated using Eq. 10 from Chanzy and Bluckler vs. volumetric water content for (A) loam and (B) silty clay loam soil under two evaporative demand conditions: U_D = 1 m/s and E_{PD} = 2 mm/d (squares) and U_D = 5 m/s and E_{PD} = 10 mm/d (diamonds).

$$E_D = 0.5 D_E t^{-1/2}$$

where the desorptivity D_E (mm/d$^{1/2}$) is assumed to be a constant for a particular soil type and t is the time (in days) from the start of S_2. Although both numerical models and observations indicate that the soil evaporation is certainly a more complicated process than the simple analytical expression given by equation ($E_D = 0.5 D_E t^{-1/2}$,) a number of field studies have shown that for S_2 conditions, reliable daily values can be obtained using Eq. above. In many of these studies for determining D_E, the integral of Eq. above is used, which yields the cumulative evaporation as a function of $t^{1/2}$

$$\sum E_D = D_E \left(t - t_O\right)^{1/2}$$

where t_O is the number of days where $E_D/E_{PD} \sim 1$ or is in S_1. In practice, observations of ΣE_D are plotted vs. $(t - t_O)^{1/2}$ and in many cases the choice of the starting point of S_2 is $t_O \approx 0$ or immediately after the soil is saturated. Evaporation rate for three drying experiments indicates that for a loam soil, t_O depends on the evaporative demand or E_{PD} with $t_O \sim 2$ days when E_{PD} is high vs. $t_O \sim 5$ days when E_{PD} is low. On the other hand, for a sandy soil, there is almost an immediate change from S_1 to S_2 conditions with $t_O \approx 1$ day. The value of t_O can significantly influence the value computed for D_E. Jackson et al. also show that for the same soil type, the value of D_E has a seasonal dependency (ranging from 4 to 8 mm/d$^{1/2}$) most likely related to the evaporative demand, which they correlate to daily average soil temperature. Values of D_E from the various studies are listed in Table 2. Brutsaert and Chen modified Eq. $E_D = 0.5D_E t^{-1/2}$ for deriving D_E by rewriting in terms of a "time-shifted" variable $T = t - t_0$ and expressing it in the form

$$E_D^{-2} = \left(\frac{2}{D_E}\right)^2 T$$

where D_E and t_0 will come from the slope and intercept. It follows that ΣE_D under S_2 will start at $T = T_0$ and not at $T = 0$, so that Eq. $\sum E_D = D_E (t - t_0)^{1/2}$ is rewritten as

$$\sum E_D = D_E \left(T^{1/2} - T_0^{1/2}\right)$$

They evaluated the effect on the derived D_E using this technique. The value of D_E using Eqs. ($E_D^{-2} = \left(\dfrac{2}{D_E}\right)^2 T$)and ($\sum E_D = D_E \left(T^{1/2} - T_0^{1/2}\right)$) was estimated to be approximately 3.3 mm/d$^{1/2}$, which is smaller than D_E values namely, 4.3–5 mm/d$^{1/2}$. This technique yields a better linear fi t to the data points that were actually under S_2 conditions.

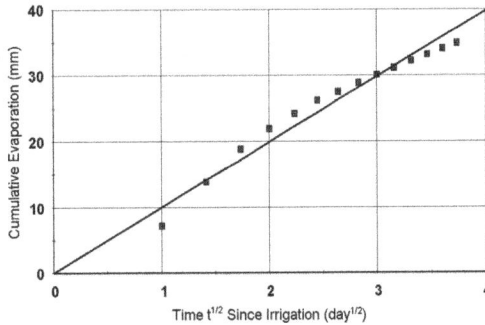

Figure: The desorptivity D_E (mm/d$^{1/2}$) estimated from a least squares linear fit to the data assuming $t_0 = 0$ (i.e., stage-two drying occurs immediately after irrigation/ precipitation).

Equations ($E_D^{-2} = \left(\dfrac{2}{D_E}\right)^2 T$) and ($\sum E_D = D_E \left(T^{1/2} - T_0^{1/2}\right)$) were used and compared to using Eq. $\sum E_D = D_E (t - t_0)^{1/2}$ with $t_0 = 0$. The plot of Eq. $\sum E_D = D_E (t - t_0)^{1/2}$ with the regression line in figure yields $D_E \sim 10$ mm/d$^{1/2}$, which is significantly larger than any previous estimates. Moreover, it is obvious from the figure that Eq. $\sum E_D = D_E (t - t_0)^{1/2}$ should not be applied with $t_0 = 0$, as this relationship is not linear over the whole drying processes. With Eq. ($E_D^{-2} = \left(\dfrac{2}{D_E}\right)^2 T$,) applied to the data, t_0 is estimated to be approximately 4.3 days, and thus a linear relationship should start at the shifted time scale $T = t - 4.3$; this means ΣE_D should start on day 5 or $T_O \approx 5 - 4.3$. With Eq. ($\sum E_D = D_E \left(T^{1/2} - T_O^{1/2} \right)$), a more realistic $D_E \approx 4.6$ mm/d1/2 is estimated for the linear portion of daily evaporation following the S_2 condition.

While this approach is relatively easy to implement operationally, D_E will likely depend on climatic factors as well as soil textural properties. However, it might be feasible to describe the main climate/seasonal effect on D_E from soil temperature observations. These might come from weather station observations or possibly from multitemporal remote sensing observations of surface temperature.

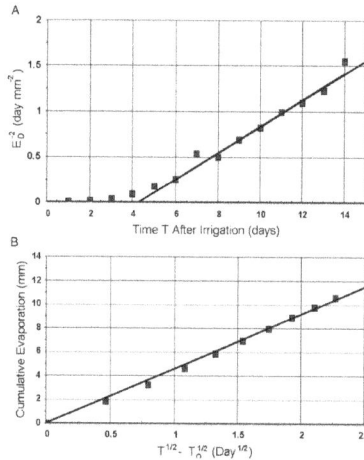

Figure: Estimation of (A) D_E and t_O with the data from figure using Eq. ($E_D^{-2} = \left(\dfrac{2}{D_E} \right)^2 T$) the resulting cumulative evaporation ΣED curve under second stage drying using equation ($\sum E_D = D_E \left(T^{1/2} - T_O^{1/2} \right)$).

The difficulty in developing a formulation for R_S, which correctly describes the water vapor transfer process in the soil, a combination method instead that involves atmospheric surface layer observations and remotely sensed surface temperature, T_{RS}. Starting with the energy balance equation

$$R_N = H + G + LE$$

where R_N is the net radiation, H the sensible heat flux, and G the soil heat flux all in W/m², and assuming the resistance to heat and water vapor transfer are similar yielding,

$$H = \rho C_P \left(\frac{T_{RS} - T_A}{R_A} \right)$$

$$LE = \frac{\rho C_P}{\gamma} \left(\frac{e_{RS} - e_A}{R_A} \right)$$

an equation of the following form can be derived

$$LE = \left(\frac{\gamma + \Delta}{\Delta}\right) LE_P - \frac{\rho C_P}{\Delta}\left(\frac{e_*(T_{RS}) - e_A}{R_A}\right)$$

where

$$LE_P = \rho C_P \left(\frac{e_*(T_A) - e_A}{R_A(\Delta + \gamma)}\right) + \Delta\left(\frac{R_N - G}{\Delta + \gamma}\right)$$

The difference $e_*(T_A) - e_A$ is commonly called the saturation vapor pressure deficit, and the value of soil surface vapor pressure e_{RS} is equal to $h_{RS} e_*(T_{RS})$ where h_{RS} is the soil surface relative humidity. Substituting Eq. ($LE_P = \rho C_P \left(\frac{e_*(T_A) - e_A}{R_A(\Delta + \gamma)}\right) + \Delta\left(\frac{R_N - G}{\Delta + \gamma}\right)$) into Eq. $LE = \left(\frac{\gamma + \Delta}{\Delta}\right) LE_P - \frac{\rho C_P}{\Delta}\left(\frac{e_*(T_{RS}) - e_A}{R_A}\right)$ yields

$$LE = R_N - G - \frac{\rho C_P}{\Delta}\left(\frac{e_*(T_{RS}) - e_*(T_A)}{R_A}\right)$$

This equation has the advantage over the above bulk resistance formulations using R_S in that there are no assumptions made concerning the saturation deficit at or near the soil surface or how to define h_{RS}. Instead, this effect is accounted for by T_{RS} because as the soil dries, T_{RS} increases and hence $e_*(T_{RS})$, which generally results in the last term on the right-hand side of Eq. $LE = R_N - G - \frac{\rho C_P}{\Delta}\left(\frac{e_*(T_{RS}) - e_*(T_A)}{R_A}\right)$ to increase, thus causing LE to decrease. In a related approach, the magnitude of LE is simply computed as a residual in the energy balance equation, Eq. $LE_P = \rho C_P \left(\frac{e_*(T_A) - e_A}{R_A(\Delta + \gamma)}\right) + \Delta\left(\frac{R_N - G}{\Delta + \gamma}\right)$, namely

$$LE = R_N - G - \rho C_P \left(\frac{T_{RS} - T_A}{R_A}\right)$$

Particularly crucial in the application of either Eq. ($LE = R_N - G - \frac{\rho C_P}{\Delta}\left(\frac{e_*(T_{RS}) - e_*(T_A)}{R_A}\right)$) or ($LE = R_N - G - \rho C_P \left(\frac{T_{RS} - T_A}{R_A}\right)$) is a reliable estimate of R_A and T_{RS}. The aerodynamic resistance R_A is typically expressed in terms of Monin-Obukhov similarity theory

$$R_A = \frac{\left\{\ln\left[\frac{z - d_o}{z_{OM}}\right] - \psi_M\right\}\left\{\ln\left[\frac{z - d_o}{z_{OS}}\right] - \psi_S\right\}}{k^2 U}$$

where z is the observation height in the surface layer (typically 2–10 m), d_o the displacement height, z_{OM} the momentum roughness length, z_{OS} the roughness length for scalars (i.e., heat and water vapor), k (~0.4) von Karman's constant, ψ_M the stability correction function for momentum, and y_S the stability correction function for scalars. Both d_o and z_{OM} are dependent on the height and density of the roughness obstacles and can be considered a constant for a given surface, while the magnitude of z_{OS} can vary for a given bare soil surface as it is also a function of the surface friction velocity. Experimental evidence suggests that existing theory with possible modification to some of the "constants" can still be used to determine z_{OS} providing acceptable estimates of H for bare soil surfaces. However, application of Eq. ($LE = R_N - G - \rho C_P \left(\dfrac{T_{RS} - T_A}{R_A} \right)$) in partial canopy cover conditions has not been successful in general because z_{OS} is not well defined in Eq.($R_A = \dfrac{\left\{ \ln \left[\dfrac{z - d_O}{z_{OM}} \right] - \psi_M \right\} \left\{ \ln \left[\dfrac{z - d_O}{z_{OS}} \right] - \psi_S \right\}}{k^2 U}$), exhibiting large scatter with the existing theory.

For this reason, estimating soil evaporation from partially vegetated surfaces using T_{RS} invariably has to involve "two-source" approaches whereby the energy exchanges from the soil and vegetated components are explicitly treated. Similarly, when using remotely sensed soil moisture for vegetated surfaces, a two-source modeling framework needs to be applied. In these two-source approaches, there is the added complication of determining aerodynamic resistances between soil and vegetated surfaces and the canopy air space. Schematically, the resistance network and corresponding flux components for two-source models. An advantage with the two-source formulation of Norman et al. is that RS is not actually needed for computing LE_S as it is solved as a residual. Nevertheless, the formulations in such parameterizations that are used (such as the aerodynamic resistance formulations) are likely to strongly influence LE_S values. Yet this two-source formulation is found to be fairly robust in separating soil and canopy contributions to evapotranspiration.

Soil Water Redistribution during Evaporation

Evaporation from the soil largely determines both water availability in terrestrial ecosystems, and the partitioning of solar radiation between sensible and latent heat. It is key to both hydrology and climate. The evaporation process is complex, involving movement and phase change of water, varying with depth and time. Following water inputs, evaporation occurs at the soil surface, controlled by atmospheric demand. As surface soil water is depleted, evaporation becomes soil-limited and shifts below the surface; nonetheless it is generally viewed as a strictly surface process. As a result, measurement methods and understanding of these near-surface phenomena have lagged behind demand for accurate data.

The rate and quantity of evaporation from a soil surface is a complicated process affected by many soil characteristics, tillage, and environmental interactions. However, it is known that energy and water availability largely dominate the process, thus on the average these broad principles can be used to estimate direct soil water evaporation. This evaporation is not to be confused with that considered as intercepted soil surface water since water from deeper that a few soil grains must be moved toward the surface to interact with eh climatic energy, thus it encounters variable resistance due to gradient flow characteristics and latent soil heat.

Soil water evaporation is represented by defining a thin (0.5-1.0 inch) upper boundary layer (evaporation layer) of the soil profile which is included in the soil profile incrementation. This upper boundary layer has all of the same functions as other layers (except no roots), plus the water is readily evaporated and limited only by PET. The lower limit of soil water content in the evaporative layer is set as just below wilting point.

Upward water movement from the second layer into the evaporation boundary layer and its evaporation is estimated by a modified Darcy equation using a reduced unsaturated conductivity rate for the current soil water content. The conductivity reduction by a small percentage represents the fact that evaporation is largely vapor flow rather than liquid and the effective conductivity is significantly less. This upward flow is obviously also dependent on the soil water content in the second and deeper soil layers. Effects such as tillage or deep soil cracking can be estimated by increasing the evaporation percentage.

This soil water evaporation routine was developed in an attempt to represent the effect of soil characteristic, and yet maintain SPAW's balance of complexity and accuracy. The programmed routine approximates the traditional observed three-stage drying process. The upper boundary layer evaporation is limited only by the PET rate (stage 1), upward movement and evaporation from a wet soil remains rapid at a decreasing rate with drying (stage 2), and evaporation from a relatively dry soil becomes very restricted (stage 3). Soil surface evaporation can accumulate to several inches over a year depending on the crop canopy, precipitation pattern and amount and PET. This soil surface water evaporation makes up the second component of actual ET.

For dry soil conditions with a partial canopy, there is some portion of the radiation energy (PET) which impinges on the soil surface but is not utilized in water evaporation. This energy heats the soil, adjacent air, and canopy, and is then reflected or absorbed and reradiated. The result is that the crop canopy has this available as a second source of PET in addition to the directly intercepted energy. To represent this effect, a linear relationship of canopy versus percent of unused energy or sensible heat absorbed by the canopy is included as shown in figure. Based only on intuitive reasoning, it was assumed that when the canopy value reaches 60 percent, all soil surface unused energy is re-captured by the canopy and it becomes a part of the potential transpiration.

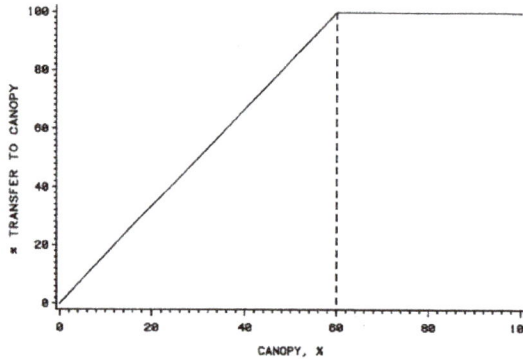

Figure: Unused soil water evaporation energy transfer to plant canopy.

Vapor Flow through Soil

Liquid water moves in the soil under saturated and unsaturated conditions.

Saturated Flow of Water

Soil is saturated with water which means that even the macrospores are full of water, and water at the surface of the soil possesses a high positive potential with respect to the water located at the lower level of the soil profile. As water moves from the state of higher positive potential to the state of lower the adjacent drier soil. But water can only flow through the pore spaces which are very irregular or even discontinuous in shape.

Saturated flow of water depends on the difference of potential which is the force causing the flow of water between different points the soil and the case with which the soil allows the ready movement of water. This latter factor is called the Hydraulic Conductivity of the soil which depends on the macrospores and the viscosity of water.

Unsaturated Flow of Water

The soil is unsaturated with water which means that the macrospores are filled up with air and the micropores are filled up with water. Water flows through micropores and the surface of the soil particles from a zone of lower negative potential to another zone of a higher negative potential.

Clay and humus attract water molecules. Water moves from moist soil where it is under a lower negative potential to dry soil. The water absorbed by dry clay possesses a very high negative potential.

Water vapour moves for high vapour pressure (generally in moist soil) to low pressure (generally in dry soil). Some vapour transfer does occur within soils. The extent of the movement by this means, however, even from one continuous macro pore to another.

There are Two Types of Water Vapour Movement

(a) Internal movement, the change from the liquid to the vapour state takes place within the soil, that is, in the soil pores, and

(b) External movement, the phenomenon occurs at the land surface and the resulting vapour is lost to the atmosphere by diffusion and convection (surface evaporation). The diffusion of water vapour from one area to another in soils does occur.

Water vapor moves through unsaturated soils in response to either thermal gradients or concentration gradients. Research has shown that increasing temperature increases the total quantity of water vapor diffusing through soil in response to concentration gradients. Even in dry soils, some moisture movement takes place in the vapor form and plays an important role in soil water regimes.

Water vapor movement in soil under isothermal conditions has also been measured. The results indicated that two processes take place in the initial period at a temperature of 20°C and a low initial water content of 6.5%; sorption of water vapor by the soil and vapor migration to the soil surface under the effect of the partial pressure gradient.

Mass Flow

Water vapour flow in a mass along with the movement of other soil gases. When the atmospheric pressure decreases soil gases including water vapour expand and move out the soil. When the atmospheric pressure is increased, soil gases and water vapour contracts and therefore gases and water vapour enter the soil.

Diffusion

Diffusion is the property by which two gases readily mix with each other when they are brought in contact with each other. Vapour pressure of water is a measure of its tendency to evaporate. Water evaporates from the moist soil and the vapour diffuses into the adjacent dry soil.

Water also evaporates from warm soil and the water vapour diffuses into the adjacent cool soil. Diffusion of water vapour due to the difference in relative humidity may balance the diffusion of water vapour due to the difference in temperature so that the net movement of water vapour may be zero as shown in figure below.

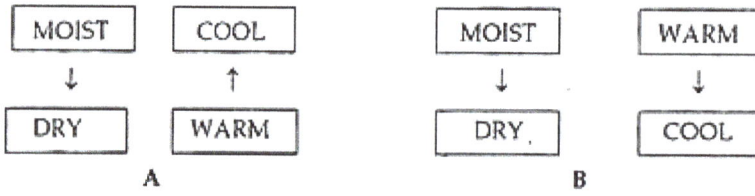

MOIST	COOL
↓	↑
DRY	WARM

A

MOIST	WARM
↓	↓
DRY	COOL

B

Figure. Movement of water vapour by diffusion

Water vapour moves from one zone to another zone only when both of them work in one direction only as shown in the figure above.

2 Theory

The absorption of water vapor by the soil causes the amount of water vapor diffusion to vary with the position in the sample. This variation can be handled mathematically by adding a sink term, α, to the equation for Fick's second law of gas diffusion in differential form to obtain:

$$\frac{\partial C}{\partial t} = D\left(\frac{\partial^2 C}{\partial x^2}\right) - \alpha$$

where C is the concentration of water vapor (g cm^{-3}), t is the time (s), D is the diffusion coefficient (cm^{2}s^{-1}), x is the distance of travel through the soil sample (cm), and α is water absorption rate per unit volume of soil (gcm^{-3}s^{-1}). From its definition, α can be calculated as water absorbed (g)/(time (s) ×soil volume (cm^3)). For a steady state condition, Eq.1 can be set equal to zero and solved to give:

$$\frac{d^2 C}{dx^2} = \frac{\alpha}{D}$$

Integration of above equation gives:

$$\frac{dC}{dx} = \alpha\frac{x}{D} + \text{constant}$$

The constant of integration can be evaluated by defining q_a as the rate of water absorption by the CaCl$_2$ (g cm^{-2}s^{-1}) and observing that at the upper surface of the sample where x=L(the sample thickness, cm). Thus

$$\frac{dC}{dx} = \alpha\frac{x}{D} - \alpha\frac{L}{D} - \frac{qa}{D}$$

Rearranging above equation, integrating, and evaluating the constant gives:

$$C = -\frac{1}{D}\left(\alpha Lx - \frac{\alpha x^2}{2} + q_a x\right) + C_w$$

where C is the water vapor concentration anywhere within the soil sample, and C_w is the value of C at x=0 (on the water side). Solving above equations for D gives:

$$D = \left(\alpha Lx - \frac{\alpha x^2}{2} + q_a x\right)\bigg/(C_w - C)$$

Evaluating Eq. above for the exit side where x=L and C=C_e

$$D = \left(q_a L + \frac{\alpha L^2}{2}\right)\bigg/(C_w - C_e)$$

Value s o f D are commonly converted to the dimensionless form D/D_0 and graphed accordingly. For this purpose, D/D_0 values for the diffusion rate of water vapor in the absence of any obstruction can be calculated from the expression $D^0 = 0.220(T/273)^{1.75}$ to be 0.2343, 0.2490, 0.2640, and 0.2795 cm²s⁻¹ for 10, 20, 30, and 40°C, respectively, where T is the absolute temperature (°K).

Transpiration

Transpiration is the loss of water from a plant in the form of water vapor. Water is absorbed by roots from the soil and transported as a liquid to the leaves via xylem. In the leaves, small pores allow water to escape as a vapor. Of all the water absorbed by plants, less than 5% remains in the plant for growth.

Reason for Transpiration of Plants

Evaporative cooling: As water evaporates or converts from a liquid to a gas at the leaf cell and atmosphere interface, energy is released. This exothermic process uses energy to break the strong hydrogen bonds between liquid water molecules; the energy used to do so is taken from the leaf and given to the water molecules that have converted to highly energetic gas molecules. These gas molecules and their associated energy are released into the atmosphere, cooling the plant.

Accessing nutrients from the soil: The water that enters the root contains dissolved nutrients vital to plant growth. It is thought that transpiration enhances nutrient uptake into plants.

Carbon dioxide entry: When a plant is transpiring, its stomata are open, allowing gas exchange between the atmosphere and the leaf. Open stomata allow water vapor to leave the leaf but also allow carbon dioxide (CO_2) to enter. Carbon dioxide is needed for photosynthesis to operate. Unfortunately, much more water leaves the leaf than CO_2 enters for three reasons:

1. H_2O molecules are smaller than CO_2 molecules and so they move to their destination faster.

2. CO_2 is only about 0.036% of the atmosphere (and rising.) so the gradient for its entry into the plant is much smaller than the gradient for H_2O moving from a hydrated leaf into a dry atmosphere.

3. CO_2 has a much longer distance to travel to reach its destination in the chloroplast from the atmosphere compared to H_2O which only has to move from the leaf cell surface to the atmosphere.

This disproportionate exchange of and H_2O leads to a paradox. The larger the stomatal opening, the easier it is for carbon dioxide to enter the leaf to drive photosynthesis; however, this large opening will also allow the leaf to lose large quantities of water and face the risk of dehydration or water-deficit stress. Plants that are able to keep their stomata slightly open, will lose fewer water molecules for every CO_2 molecule that enters and thus will have greater water use efficiency (water lost/ CO_2 gained). Plants with greater water use efficiencies are better able to withstand periods when water in the soil is low.

Water uptake: Although only less than 5% of the water taken up by roots remains in the plant, that water is vital for plant structure and function. The water is important for driving biochemical processes, but also it creates turgor so that the plant can stand without bones.

How fast does water move through plants? Transpiration rates depend on two major factors: 1) the driving force for water movement from the soil to the atmosphere and 2) the resistances to water movement in the plant.

Driving force: The driving force for transpiration is the difference in water potential between the soil and the atmosphere surrounding the plant. This difference creates a gradient, forcing water to move toward areas with less water. The drier the air around the plant, the greater the driving force is for water to move through the plant and the faster the transpiration rate.

Resistances: There are three major resistances to the movement of water out of a leaf: cuticle resistance, stomata resistance and boundary layer resistance. These resistances slow water movement. The greater any individual resistance is to water movement, the slower the transpiration rate.

A simple equation describing how these factors alter transpiration is:

$$\text{Transpiration} = \frac{[\text{Water potential}_{(leaf)}] - [\text{Water potential}_{(atmosphere)}]}{\text{Resistance}}$$

The units for this equation are mols of water lost per leaf area per time ($mol/cm^2/s$). This equation makes predicting rates of transpiration easy. For example, any time the numerator (the value for the driving force) is increased, the rate of transpiration becomes faster and vice versa. Similarly, if the denominator (the value for resistance) increases, this means there is greater resistance and thus, slower transpiration.

References

- Soil, movement-of-water-and-vapour-in-the-soil-3534: soilmanagementindia.com, Retrieved 11 July 2018

- NSF, EAR-0809656: grantome.com, Retrieved 29 June 2018

- Water-Vapor-Diffusion-Through-Soil-as-Affected-by-Temperature-and-Aggregate-Size-225254850: researchgate.net, Retrieved 25 March 2018

- Transpiration, feature, webfeat, vis-2005: sciencemag.org, Retrieved 11 May 2018

- 3-main-types-of-movement-of-water-within-soil-1176: soilmanagementindia.com, Retrieved 29 April 2018

Solute Transport in Soil

In order to study the phenomenon of solute transport in the soil, it is necessary to understand the processes by which nutrients are transported in the soil. This chapter has been written to provide a comprehensive understanding of this area of study.

Solutes found in soils may include nutrients, pesticides, salts or other naturally occurring or applied chemicals. In the soil environment, many of these solutes are beneficial as they provide plants and soil organisms with food and pest resistance; yet, the movement of solutes off -site to surface and ground water sources can have substantial agronomic, environmental and economic consequences. For example, water sources contaminated by certain nutrients and pesticides can be rendered unsafe for human and animal consumption, and may be toxic to aquatic organisms. The costs associated with off -site solute movement from an over application of agrichemicals, ineffective treatment of targeted pests (weeds, insects or diseases) or the remediation of contaminated water sources can be quite high. Therefore, understanding how solutes move in soil and learning methods to minimize off -site contamination are important for using chemicals effectively and protecting water quality.

- Advection: the spreading of a pollutant by groundwater flow.

- Diffusion: the spreading of a species dissolved in the water phase by the Brownian motion of the ions (molecules).

- Dispersion: the spreading of a species dissolved in the water phase by local variations in the water velocity.

- Adsorption/desorption: interaction of species dissolved in the water phase with the solid matrix. This process can be physically based or chemically based, reversible or irreversible.

- Chemical reactions: reactions of species dissolved in the water phase with other species, resulting in the occurrence of different species altogether.

- Biodegradation: the degradation of species dissolved in the water phase by bacteria.

- Radioactive decay: the degradation of species by radioactivity.

Concentrations of species in the water phase Ci (including pure water itself) are defined as the mass of the species per unit volume: kg/m³, g/l, mg/l, etc.

The density of a multi-component fluid, consisting of N components, is then given as:

$$\rho = \sum_{i=1}^{N} C_i$$

(1)

Mass fractions ω of the components (mass per unit of mass: kg/kg, g/g, etc.) are defined as:

$$\omega_i = \frac{C_i}{\rho} \ such \ that \ \sum_{i=1}^{N} \omega_i = 1$$

(2)

For dilute solutions (tracer concentrations) all mass fractions ω_i <<1, except for the pure water. This means that the density of the fluid is close to the density of pure water, and can be assumed to be constant.

Water density is a function of pressure, temperature and composition. This last dependence is only important at high concentrations. E.g. in case of seawater intrusion, or in deep saline aquifers which are sometimes used to store waste or to produce energy. In these deep aquifers salt concentrations can be as high as 300 g/l, resulting in a water density of 1200 g/l (giving a salt mass fraction of 0.25). Water density fluctuations will also play a role in the subsurface storage of heat.

Water viscosity is a function of pressure, temperature and composition. This influences the hydraulic conductivity. The dependence on the temperature is by far the most important. Hence, this dependence must be taken into account in the analysis of subsurface storage of heat.

Groundwater Flow

Groundwater flow is described by Darcy's law. Darcy's law is in principle the form of the momentum balance (Navier-Stokes equation), averaged over a large number of pores. It also follows from a balance of forces on water flowing through a porous medium.

Consider flow in the z-direction. The net forces (positive upward) working on a body of water with dimensions Δx, Δy and Δz are:

Pressure forces: $(p(z)-p(z+\Delta z))\Delta x \Delta y$

Gravity forces: $-\rho g \Delta x \Delta y \Delta z$

Friction forces: $-R'q_z \Delta x \Delta y \Delta z$

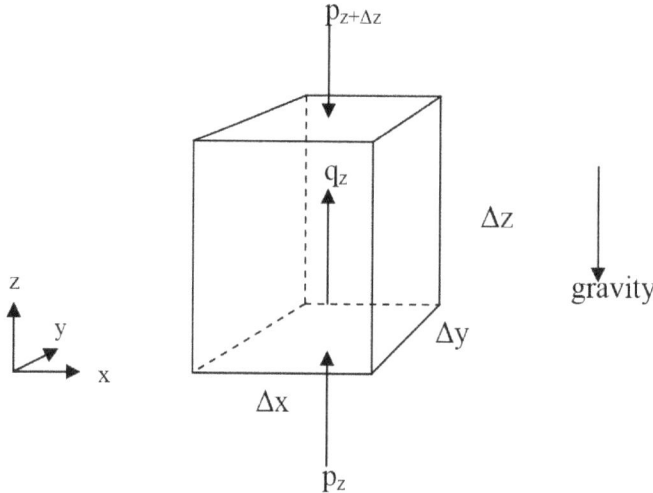

Figure: Balance of forces on water in a porous medium

where R' is the resistance factor and q_z is the specific discharge (Darcy velocity) in the z direction.

For the pressure forces, we can make the following approximation by using a first order Taylor series expansion:

$$p_z - p_{z+\Delta z} = p_z - \left(p_z + \frac{\partial p}{\partial z} \Delta z \right) = -\frac{\partial p}{\partial z} \Delta z \tag{3}$$

Then setting up the force balance, we find:

$$q_z = -\frac{1}{R'} \left(\frac{\partial p}{\partial z} + \rho g \right) = -\frac{\kappa_z}{\mu} \left(\frac{\partial p}{\partial z} + \rho g \right) \tag{4}$$

where we have assumed that the resistance factor R' is proportional to the liquid viscosity μ. κ_z is the intrinsic permeability in the z-direction (L^2), which is assumed to be a property of the porous medium. The intrinsic permeability of a porous medium is largely determined by the pore sizes and shapes. A strong correlation between permeability and porosity exists. Similar expressions can be obtained for the flow in x and y direction:

$$q_x = \frac{\kappa_x}{\mu} \frac{\partial p}{\partial x} \qquad q_y = -\frac{\kappa_y}{\mu} \frac{\partial p}{\partial y} \tag{5}$$

Basic assumptions in this derivation are that the acceleration of the water can be neglected, and that the friction forces are linear dependent on the velocity.

The latter is not always true (especially at high water velocities, e.g. close to an abstraction or infiltration well), in which case Darcy's law is not valid, but should be replaced by Forchheimer's equation:

$$q_x + \beta q_x^2 = -\frac{\kappa_x}{\mu}\frac{\partial p}{\partial x} \qquad q_y + \beta q_y^2 = -\frac{\kappa_y}{\mu}\frac{\partial p}{\partial y} \qquad q_z + \beta q_z^2 = -\frac{\kappa_z}{\mu}\left(\frac{\partial p}{\partial z} + \rho g\right) \tag{6}$$

where β is again a property of the porous medium.

Define a piezometric head h as:

$$h = \frac{p}{\rho g} + z \quad or \quad p = \rho g(h - z) \tag{7}$$

Basically, the piezometric head consists of a pressure head $p/\rho g$ and the vertical position z with respect to the reference level. It is the position of the top of the water column in an observation well with respect to the reference level (usually mean sea level). This is different from unsaturated flow, which is formulated in terms of the pressure head. Substitution of equation above in equation $q_z = -\frac{1}{R'}\left(\frac{\partial p}{\partial z} + \rho g\right) = -\frac{\kappa_z}{\mu}\left(\frac{\partial p}{\partial z} + \rho g\right)$, assuming that the density ρ is constant then gives Darcy's law in terms of the groundwater head h:

$$q_z = -\frac{\kappa_z}{\mu}\left(\frac{\partial p}{\partial z} + \rho g\right) = -\frac{\kappa_z}{\mu}\left(\rho g\left(\frac{\partial h}{\partial z} - 1\right)\rho g\right) = -\frac{\kappa_z \rho g}{\mu}\frac{\partial h}{\partial z} \tag{8}$$

and similar expressions can be obtained for q_x and q_y.

Consequently, if the density ρ and the viscosity μ are constant, we can define hydraulic conductivities as:

$$K_x = \frac{\kappa_x \rho g}{\mu} \qquad q_x = -k_x \frac{\partial h}{\partial x} \tag{9}$$

and the same for the y and z direction. This shows that the hydraulic conductivity k (L/T) is dependent on the fluid properties.

The groundwater flow equation follows from a mass balance for the complete water phase (including all dissolved species). Consider the element as depicted in figure below with dimensions Δx, Δy and Δz. The net mass influx over a period Δt in the x-direction is given by:

$$\left(\rho q_x(x) - \rho q_x(x + \Delta x)\right)\Delta y \Delta z \Delta t \approx \left(\rho q_x(x) - \left(\rho q_x(x) + \frac{\partial(\rho q_x)}{\partial x}\Delta x\right)\right)\Delta y \Delta z \Delta t$$

$$= \frac{\partial(\rho q_x)}{\partial x}\Delta x \Delta y \Delta z \Delta t \tag{10}$$

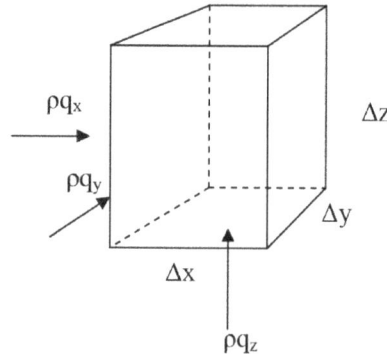
Figure: Water balance in a porous medium

A similar expression can be obtained for the net mass influx in the y and z directions. The change in the total mass in the element is given by:

$$\left(n\rho(t+\Delta t)-n\rho(t)\right)\Delta x\Delta y\Delta z \approx \frac{\partial}{\partial t}(n\rho)\Delta x\Delta y\Delta z\Delta t \tag{11}$$

where n is the porosity.

Equating the net mass influx and the change in mass gives the mass balance equation for the liquid phase:

$$\frac{\partial}{\partial t}(n\rho)+\frac{\partial}{\partial x}(\rho q_x)+\frac{\partial}{\partial y}(\rho q_y)+\frac{\partial}{\partial z}(\rho q_z)=0 \tag{12}$$

For situations with varying fluid properties (salt water intrusion, storage of heat, etc.) this equation together with the pressure formulation of Darcy's law (equations $q_z=-\frac{1}{R'}\left(\frac{\partial p}{\partial z}+\rho g\right)=-\frac{\kappa_z}{\mu}\left(\frac{\partial p}{\partial z}+\rho g\right)$ and ($q_x=\frac{\kappa_x}{\mu}\frac{\partial p}{\partial x}$ $q_y=-\frac{\kappa_y}{\mu}\frac{\partial p}{\partial y}$) should be used. Note, that the flow equation in that case is non-linear. Note also, that in these cases, even though a piezometric head can be defined, it will not be the driving force for groundwater flow.

If the density of the liquid ρ and the porosity n is assumed to be dependent on the pressure p only, the time derivative in the mass balance equation can be written as:

$$\frac{\partial}{\partial t}(n\rho)=\left[\rho\frac{\partial n}{\partial p}+n\frac{\partial \rho}{\partial p}\right]\frac{\partial p}{\partial t}=\rho g\left[\rho\frac{\partial n}{\partial p}+n\frac{\partial \rho}{\partial p}\right]\frac{\partial h}{\partial t}=\rho S_s\frac{\partial h}{\partial t} \tag{13}$$

where S_s is the specific storage. Combining this equation with the piezometric head formulation of Darcy's law equations $q_z=-\frac{\kappa_z}{\mu}\left(\frac{\partial p}{\partial z}+\rho g\right)=-\frac{\kappa_z}{\mu}\left(\rho g\left(\frac{\partial h}{\partial z}-1\right)\rho g\right)=-\frac{\kappa_z\rho g}{\mu}\frac{\partial h}{\partial z}$

and $K_x = \dfrac{\kappa_x \rho g}{\mu}$ $\quad q_x = -k_x \dfrac{\partial h}{\partial x}$ and division by the (constant) density gives the well-known groundwater flow equation:

$$\rho S_s \frac{\partial h}{\partial t} - \frac{\partial}{\partial x}\left(k_x \frac{\partial h}{\partial x} \right) - \frac{\partial}{\partial y}\left(k_y \frac{\partial h}{\partial y} \right) - \frac{\partial}{\partial z}\left(k_z \frac{\partial h}{\partial z} \right) = 0 \tag{14}$$

Note, that the average pore water velocity   is different from the specific discharge q:

$$\upsilon = q / n.$$

Simplified Description of Processes in Reactive Transport

General

Similar to the water balance, we can derive a general form for the mass balance of a dissolved component in groundwater. Assume that the mass fluxes in x, y and z-directions are given by F_x, F_y and F_z (M/L²T) respectively.

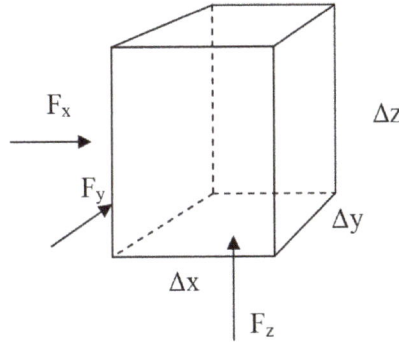

Figure: General mass balance for a dissolved component in a porous medium

The net mass influx in the x-direction over a period Δt is then given by:

$$\left(F_x(x) - F_x(x + \Delta x) \right) \Delta y \Delta z \Delta t \approx \left[F_x(x) - \left(F_x(x) + \frac{\partial F_x}{\partial x} \Delta x \right) \right] \Delta y \Delta z \Delta t$$

$$= -\frac{\partial F_x}{\partial x} \Delta x \Delta y \Delta z \Delta t \tag{15}$$

and similar expressions can be obtained for the net mass influx in the y and z-directions.

The change in mass of the component in the element over a period Δt is given by:

$$\left(nC(t + \Delta t) - nC(t) \right) \Delta x \Delta y \Delta z \approx \frac{\partial}{\partial t}(nC) \Delta x \Delta y \Delta z \Delta t \tag{16}$$

Due to the different processes occurring, mass of a component can be produced or lost in a period, e.g. because of adsorption/desorption, chemical reactions, decay, etc. The loss of mass due to these processes per unit volume and unit time will be indicated by I (M/L³T). Note, that I can be either positive (loss of mass) or negative (gain of mass). Combining the different terms then gives the following general mass balance equation:

$$\frac{\partial}{\partial t}(nC) + \frac{\partial}{\partial x}(F_x) + \frac{\partial}{\partial y}(F_y) + \frac{\partial}{\partial z}(F_z) + I = 0 \tag{17}$$

In the following, the mass fluxes and/or the mass production associated with the different processes playing a role will be given. For the time being, simplified (linear) expressions will be given, which will result in a mass balance equation in the form of the classical Advection Dispersion (or Convection Dispersion) equation, CDE. Later, more complicated expressions will be covered.

Advection

Advection (or convection) is the transport of dissolved components by flowing groundwater. The mass transport per unit area of porous medium of a dissolved component by flowing groundwater is given by:

$$F_x = q_x C \tag{18}$$

where F_x is the mass flux of the component in the x-direction (M/L²T), q_x is the specific discharge of water (Darcy velocity) in the x-direction (L³/L²T) and C is the concentration of the component in the water phase (M/L³). No mass is produced or lost, hence, $I = 0$.

The underlying assumption is that the average velocity of the ions or molecules of the dissolved substance is the same as the average water velocity: if we move one liter of water over a certain distance, also all chemicals in that liter will have moved that distance. In most cases, this will be true, but there are exceptions. These exceptions occur e.g. when the molecules of the dissolved substance are very large (colloids, virus). If we consider the flow of water in a capillary, the water velocity v at a distance r from the centre is given by:

$$v = 2v_{avg}\left(1 + \frac{r^2}{r_0^2}\right) \tag{19}$$

where v_{avg} is the average water velocity and r_0 is the radius of the capillary. For large molecules, only part of the capillary is available for transport. That can be caused by either the size of the molecules or by the electrical charges on the surface. As a result, the average velocity of such particles in a capillary will exceed the average velocity of the water itself. If the radius of such particles is given by r_c, it can easily be inferred that the average velocity of the particles compared to the average water velocity of the water is given by:

$$\frac{\upsilon_C}{\upsilon_{avg}} = 1 + 2\alpha + \alpha^2 \qquad\qquad \alpha = \frac{r_C}{r_0} \tag{20}$$

(find the average water velocity by integrating the water velocity from 0 to r_0, and find the average particle velocity by integrating the water velocity from 0 to $r_0 - r_c$). For instance, for particles with a size of 20% of the capillary diameter, the average velocity is some 40% larger than the water velocity. These effects have been observed in virus and colloid transport.

If only advective transport takes place, the mass balance for a component follows from (17) and (18):

$$\frac{\partial}{\partial t}(nC) + \frac{\partial}{\partial x}(q_x C) + \frac{\partial}{\partial y}(q_y C) + \frac{\partial}{\partial x}(q_z C) = 0 \tag{21}$$

Now consider the 1-dimensional mass balance equation with constant porosity n and constant specific discharge q. Division by n then gives the following equation:

$$\frac{\partial}{\partial t}(C) + \frac{q}{n}\frac{\partial}{\partial x}(C) = \frac{\partial}{\partial t}(C) + \upsilon\frac{\partial}{\partial x}(C) = \frac{DC}{Dt} = 0 \tag{22}$$

where DC/Dt is the material derivative, i.e. the change in concentration when moving along with a water particle. Since this derivative is zero, advective transport only results in a displacement of the initial concentration distribution by υ_t, where t is the elapsed time. This is also true in 3 dimensions.

Equation (22) can be written in dimensionless form by defining the following dimensionless variables:

$$C_d = \frac{C}{C_r} \qquad t_d = \frac{t}{t_r} \qquad x_d = \frac{x}{L_r} \tag{23}$$

where C_r, t_r and L_r are reference or characteristic values for the system considered. Substitution of these dimensionless variables in the mass balance equation (22) gives:

$$\frac{\partial C_d}{\partial t_d} + \frac{\upsilon t_r}{L_r}\frac{\partial C_d}{\partial x_d} = 0 \tag{24}$$

We can now choose any one of the characteristic values t_r or L_r such that the coefficient in front of the spatial derivative in (24) is 1. This means that for a given characteristic time t_r, the characteristic length is given by υt_r, while for a given characteristic length L_r, the characteristic time is given by L_r/υ. These characteristic values obviously are related to respectively the travel distance and the travel time of water particles.

The specific discharge that is required to quantify the advective fluxes follows from the mass balance equation for the water phase in combination with Darcy's law. Basically, this means that (local) information about the value of the permeability (or hydraulic conductivity) is required.

Diffusion

Diffusion is the spreading of a component dissolved in the water phase by the Brownian motion of the molecules/ions. In open water, the mass flux due to diffusion is given by Fick's first law:

$$F_x = -D_m \frac{\partial C}{\partial x}$$

(25)

where D_m is the molecular diffusion coefficient (L^2/T), which is typical for the component considered. In a porous medium, the mass flux due to diffusion is given by a similar expression:

$$F_x = -n \frac{D_m}{\tau} \frac{\partial C}{\partial x} = -n D_{eff} \frac{\partial C}{\partial x}$$

(26)

The porosity n enters the equation to account for the area that is effectively available for mass transport. τ is the tortuosity of the porous medium (-), which accounts for the fact that the length of the path molecules or ions have to take in a porous medium to travel from one position to another is larger than the distance between these positions. For normal porous media, τ has a value in the order of 1.6 to 1.7. No mass is produced or lost due to diffusion, hence I=0.

In case only diffusion occurs, the mass balance equation reads:

$$\frac{\partial}{\partial t}(nC) - \frac{\partial}{\partial x}\left(nD_{eff}\frac{\partial C}{\partial x}\right) - \frac{\partial}{\partial y}\left(nD_{eff}\frac{\partial C}{\partial y}\right) - \frac{\partial}{\partial z}\left(nD_{eff}\frac{\partial C}{\partial z}\right) = 0$$

(27)

Now consider a 1-dimensional mass balance equation with constant porosity n and diffusion coefficient D_{eff}:

$$\frac{\partial C}{\partial t} - D_{eff}\frac{\partial^2 C}{\partial x^2} = 0$$

(28)

For this equation, numerous analytical solutions dependent on the boundary conditions are known, both in Cartesian and radial coordinate systems. The steady state solution in a Cartesian coordinate system is:

$$C = Ax + B$$

(29)

where A and B are determined by the boundary conditions. Note, that this solution is not dependent on the effective diffusion coefficient.

Equation (28) can be written in dimensionless for by defining appropriate characteristic values for the concentration and the time and length scales:

$$\frac{\partial C_d}{\partial t_d} - \frac{D_{eff} t_r}{L_r^2} \frac{\partial^2 C_d}{\partial x_d^2} = 0 \tag{30}$$

Setting the coefficient in front of the spatial derivative to 1, characteristic values for the time and the length are found. For a given characteristic time t_r, the characteristic length is given by $L_r = \sqrt{(D_{eff} t_r)}$, and for a given characteristic length L_r, the characteristic time is given by $t_r = L_r^2 / D_{eff}$.

In general, molecular diffusion will not play an important role in porous media transport, unless the groundwater velocities are very small (which is e.g. the case for transport through very low permeable clays).

Measurement of effective diffusion coefficients in a porous medium is usually done in the laboratory by performing time dependent experiments.

Dispersion

Dispersion is the spreading of a dissolved component due to local variations in the ground- water velocity. In general, we distinguish mechanical and hydrodynamic dispersion.

Mechanical dispersion takes place on the pore scale, and is caused by velocity variations across the cross section of the capillaries (or pores). Usually the groundwater velocities are so small, as are the pore diameters, that molecular diffusion is fast enough to balance concentration differences in the direction perpendicular to the flow (i.e. across the pores).

Hydrodynamic dispersion is the sum of molecular diffusion and mechanical dispersion. It usually occurs on a larger scale than a single pore, and is caused by all variations in the average groundwater velocity (i.e., averaged over a large number of pores) that we did not account for explicitly, including diffusion. Thus, if we consider layers with different values of the hydraulic conductivity (or permeability), this variation does not necessarily give rise to hydrodynamic dispersion. However, it is clear that such variation certainly may lead to variation of the rate of displacement of chemicals and to true mixing, if it combines with diffusion. This is commonly called macro or mega dispersion.

For both mechanic and hydrodynamic dispersion, the mass fluxes are assumed to be given by the following form of Fick's first law:

$$F_x = -D_{xx} \frac{\partial C}{\partial x} - D_{xy} \frac{\partial C}{\partial y} - D_{xz} \frac{\partial C}{\partial z} \tag{31}$$

where D_{xx}, D_{xy} and D_{xz} are elements of the dispersion tensor (L^2/T). Similar expressions are valid for the mass fluxes in y and z-direction. The dispersion tensor is symmetric, and consists of 6 different numbers, D_{xx}, $D_{xy} = D_{yx}$, $D_{xz} = D_{zx}$, D_{yy}, $D_{yz} = D_{zy}$ and D_{zz}. The elements of the dispersion tensor are dependent on the groundwater velocity v, such that the dispersion coefficients in the direction of the flow and perpendicular to the flow are given by:

$$D(//v) = \alpha_l |v| \qquad\qquad D(\perp v) = \alpha_t |v| \tag{32}$$

where α_l and α_t are the longitudinal and transversal dispersivities (L) respectively. These are assumed to be properties of the porous medium, and indicate the size of heterogeneities in the system that is not accounted for by variations in the (average) groundwater velocity. Because mass transport by hydrodynamic dispersion and by molecular diffusion is described by the same law, they are usually combined.

In a fully 3-dimensional system, with velocity components v_x, v_y and v_z respectively, the elements of the hydrodynamic dispersion (including molecular diffusion) are given by:

$$D_{xx} = D_{eff} + \alpha_t |v| + (\alpha_l - \alpha_t) \frac{v_x^2}{|v|}$$

$$D_{yy} = D_{eff} + \alpha_t |v| + (\alpha_l - \alpha_t) \frac{v_y^2}{|v|}$$

$$D_{zz} = D_{eff} + \alpha_t |v| + (\alpha_l - \alpha_t) \frac{v_z^2}{|v|}$$

$$D_{xz} = D_{zx} = (\alpha_l - \alpha_t) \frac{v_x v_z}{|v|}$$

$$D_{yz} = D_{zy} = (\alpha_l - \alpha_t) \frac{v_y v_z}{|v|}$$

$$D_{xy} = D_{yx} = (\alpha_l - \alpha_t) \frac{v_x v_y}{|v|} \tag{33}$$

Because hydrodynamic dispersion occurs only in combination with groundwater flow, a mass balance for a component follows from the combination of mass fluxes as defined by equations (18) and (31), with the elements of the dispersion tensor given by equation (33). Using a short hand notation, this mass balance is given by:

$$\frac{\partial}{\partial t}(nC) + \nabla \cdot (qC) - \nabla \cdot (nD \cdot \nabla C) = 0 \tag{34}$$

Now, consider a one-dimensional system with constant porosity, velocity and dispersion coefficient:

$$\frac{\partial C}{\partial t} + v\frac{\partial C}{\partial x} - D\frac{\partial^2 C}{\partial x^2} = 0 \tag{35}$$

Making this equation dimensionless by choosing appropriate characteristic values for the concentration, time and length gives:

$$\frac{\partial C_d}{\partial t_d} + \frac{v t_r}{L_r}\frac{\partial C_d}{\partial x_d} - \frac{D t_r}{L_r^2}\frac{\partial^2 C_d}{\partial x_d^2} = 0 \tag{36}$$

Choosing the characteristic time $t_r = L_r/v$ then gives the following dimensionless mass balance equation:

$$\frac{\partial C_d}{\partial t_d} + \frac{\partial C_d}{\partial x_d} - \frac{1}{Pe}\frac{\partial^2 C_d}{\partial x_d^2} = 0 \qquad\qquad Pe = \frac{v L_r}{D} \tag{37}$$

where Pe is the Peclet number. This number is characteristic for the ratio of advective transport and dispersive transport. The solution of (57) depends only on Pe, and large Peclet numbers indicate that advection dominates; small Peclet numbers indicate that dispersion dominates.

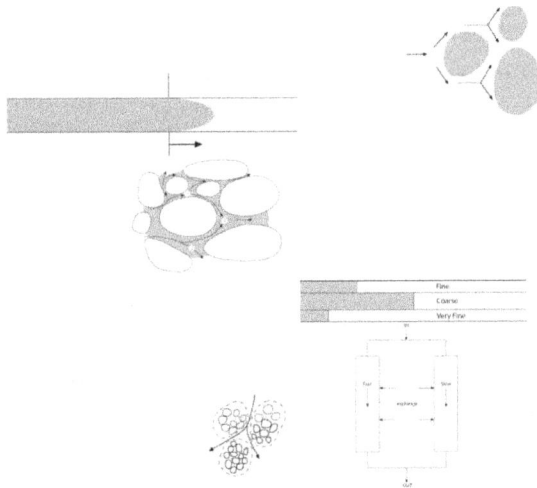

Figure: Illustration of different mechanisms of dispersion.

Mixing occurs due to velocity variations within pores, and between pores, in combination with diffusional mixing at locations where water flows with different concentrations meet. In addition, flow velocities may be aligned in the mean flow direction, but it can also have components that are at an angle with this direction. Also larger scale variations in flow velocity, due to aggregation and layering, may lead to enhanced mixing.

For mechanical dispersion, the longitudinal dispersivity α_l is in the order of the pore sizes (mm scale). For hydrodynamic dispersion, the longitudinal dispersivity is dependent on the scale of the problem. For laboratory experiments in columns, values of less than 1 mm to values larger than 1 cm have been reported. For field scale experiments values larger than 10 m have been reported, dependent on the size of the experiment and the heterogeneity of the aquifer in which the experiment was performed.

A cautioning remark is needed with regard to such large dispersivities, as such large values cannot possibly be due to complete true mixing of water of different composition in porous media. Such large values are commonly obtained based on several methodological complications: (i) the equipment used to measure 'local' concentrations (e.g. observation wells, geophysical methods) are themselves responsible for mixing, and (ii) modelling with large spatiotemporal discretization in view of computational efficiency may lead to numerical mixing, and (iii) dispersivities may be 'fitted' using a relatively simple transport equation (e.g. a one dimensional version of the transport equation), which leads to artifacts.

Based on field experiments, an empirical relation for the longitudinal dispersivity has proposed as:

$$a_1 \approx 0.0175L^{1.46} \tag{38}$$

where L is a characteristic length for the domain of interest. For large scales (large values of L), an upper bound for α_l is reached. Note, that relation (38) is based on the evaluation of a number of field experiments, and gives an estimate of the dispersivities only. Also note, that these estimates are based on the assumption that the aquifer permeability is homogeneous over the domain of interest.

As a guideline, the transversal dispersivity is commonly assumed to be 5 to 10% of the longitudinal dispersivity.

Adsorption/Desorption

Adsorption/desorption creates a sink/source term for a component in the water phase. Two processes take place at the same time: molecules/ions will attach to the solid material, and attached molecules/ions can be released from the solid to the water phase. For the time being, we will adopt a linear description of the process, corresponding with case A of figure:

$$I = k_a C - k_d C_s = k_d \left(\frac{k_a}{k_d} C - C_s \right) = k_d \left(k_d C - C_s \right) \tag{39}$$

where k_a and k_d are the attachment (1/T) and detachment (M/L³T) coefficients respectively, and C_s is the concentration of the component adsorbed (M/M). k_a and k_d have different units because of the different units for C and C_s respectively. In this formulation, no equilibrium has been assumed. K_d is the distribution coefficient. In case of equilibrium, the expression in (39) between brackets is 0, hence:

$$C_s = K_d C \tag{40}$$

which defines a linear adsorption isotherm. In principle, such a relation is only valid if the concentrations are very low, and if equilibrium between the water phase and the solid material exists.

If advection, dispersion and linear adsorption/desorption occur, the mass balance equation for a component in the water phase can be given as:

$$\frac{\partial}{\partial t}(nC) + \nabla \cdot (qC) - \nabla \cdot (nD \cdot \nabla C) + k_a C - k_d C_s = 0 \tag{41}$$

Note, that this equation is not based on the assumption of local equilibrium. Also, two unknowns are present in this equation: C and Cs . The other equation required to solve for the concentrations is given by a mass balance equation for the adsorbed component:

$$\frac{\partial}{\partial t}((1-n)\rho_s C_s) - k_a C + k_d C_s = 0 \tag{42}$$

where ρ_s is the density of the solid material. Basically, this mass balance equation is comparable to equation (41), because the adsorbed component is not transported (no advection and dispersion), and that the source term due to adsorption/desorption has the opposite sign.

Adding equations (41) and (42) gives the total mass balance:

$$\frac{\partial}{\partial t}(nC) + \frac{\partial}{\partial t}((1-n)\rho_s C_s) + \nabla \cdot (qC) - \nabla \cdot (nD \cdot \nabla C) = 0 \tag{43}$$

If we now assume equilibrium, equation (40) can be used to eliminate C_s from equation (43):

$$\frac{\partial}{\partial t}(nRC) + \nabla \cdot (qC) - \nabla \cdot (nD \cdot \nabla C) = 0 \quad R = 1 + \frac{1-n}{n}\rho_s K_d \tag{44}$$

where R is the retardation factor. Note, that for non-reactive solutes (no adsorption/desorption) $R = 1$. It is clear from this equation that the retardation factor is only found in the time-derivative term. For this reason, it got its name, as this factor R implies that both ad/convection and dispersion are R times slower: they are retarded by a factor R.

Now consider a one-dimensional form of this mass balance, with constant porosity, velocity, dispersion coefficient and retardation factor:

$$\frac{\partial C}{\partial t} + \frac{v}{R}\frac{\partial C}{\partial x} - \frac{D}{R}\frac{\partial^2 C}{\partial x^2} = 0 \tag{45}$$

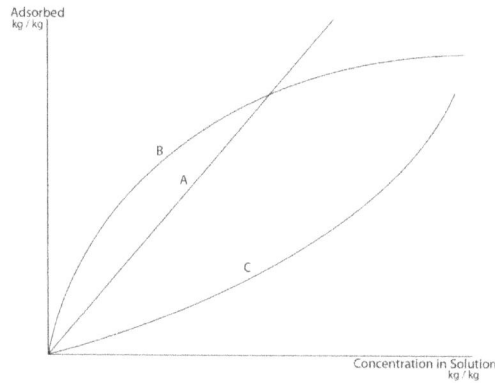

Figure: linear (A) and nonlinear adsorption, with B similar to Freundlich and
Langmuir type equations and C resembling precipitation controlled
reactions, if the nonlinearity is much more distinct than in this figure.

Note that this equation is identical to equation (35), the mass balance equation for a
nonreactive component, with both the velocity v and the dispersion D scaled by a factor
R. If the molecular diffusion can be neglected, the dispersion coefficient is proportional
to the velocity v, and equation (45) will give the same results as the mass balance equation for a non-reactive component with a velocity that is decreased by a factor R.

Now consider the one-dimensional form of equation (41) (non-equilibrium) with constant porosity, velocity, dispersion coefficient and attachment and detachment constants. This equation can be made dimensionless by choosing appropriate values for
the characteristic concentration, time and length. If (as done before) we define the
characteristic time $t_r = L_r/v$, the non-dimensional equation is given by:

$$\frac{\partial C_d}{\partial t_d} + \frac{\partial C_d}{\partial x_d} - \frac{1}{Pe}\frac{\partial^2 C}{\partial x^2} + \frac{k_a L_r}{nv}C_d - \frac{k_a L_r}{nv}C_{sd} = 0 \tag{46}$$

The last two coefficients in this equation are two forms of the dimensionless Damkohler
number. This number gives the ratio of the groundwater travel time and the time required to reach equilibrium. Large Damkohler numbers indicate that the assumption
of local equilibrium is appropriate, while small Damkohler numbers indicate that ad-
sorption/desorption should be described as a non-equilibrium process.

Measurement of the adsorption distribution coefficient K_d is commonly done in a laboratory batch experiment. A soil sample is mixed with water that contains a dissolved
component at a certain concentration. This mixture is stirred gently for a long time in
order to assure that equilibrium between the water phase and the soil is established.
From a measurement of the resulting concentration in the water phase at equilibrium,
the amount adsorbed can be determined, and the distribution coefficient calculated:

Assume we have a mass of soil M_s. we add a volume of water V which has dissolved in it a component at concentration C_i. At equilibrium, the concentration in the water phase is measured as C_{eq}. The amount of mass of the component added to the system is VC_i. At equilibrium, the total mass of the component in the water phase is VC_{eq}. Consequently, the total mass adsorbed is $V(C_i - C_{eq})$, and the concentration of adsorbed component is $C_s = V(C_i - C_{eq})/Ms$ while the concentration in the water phase is C_{eq}. The value of the distribution coefficient follows directly from equation (40).

Measurement of the attachment and detachment coefficients can be done in batch experiments by measuring the concentration in the water phase as a function of time.

Interaction of solutes do not only take place with the solid material, but can also exist with colloïdal particles (e.g. natural organic material), which in itself are mobile. Consequently, a competition between adsorption on the solid matrix and adsorption on colloïds may occur. This may lead to an enhanced transport of species (e.g. heavy metals) that may otherwise be considered to be highly retarded.

Decay

For the simplified description, we will assume that the decay due to chemical reactions, biological activity and/or radioactivity is given by a first order expression:

$$I = n\lambda C \tag{47}$$

where λ is the decay/degradation constant. λ is related to the half life $t_{1/2}$ of the component by:

$$\lambda = \frac{In(2)}{t_{1/2}} \tag{48}$$

The half-life $t_{1/2}$ is commonly measured in batch experiments by mixing a sample of soil material with water which has the component dissolved in it. Measuring the concentration in the water phase as a function of time will give an estimate of the decay. In these experiments, adsorption/ desorption should also be taken into account. The process of first order decay/degradation is of great importance for much of the transport theory. As may be already apparent, radionuclides decay proportional to the total decaying mass present. For instance the recent tsunami accident with the Fukushima nuclear plant in Japan may have resulted in soil contamination with radionuclides, where the decay rate determines the period for which radiation problems may be acute. Likewise, the Chernobyl melt down resulted in continental scale contamination with radionuclides by different elements, that move towards groundwater with different rates, and different degradation rates. To appreciate the hazard for life, the rate of downward movement of chemicals in relation with the decay rate of hazardous radiation is a typical transport problem.

The first order degradation rate law is also the most commonly used rate law for describing the degradation of contaminants such as pesticides, nutrient chemicals such as nitrate, contaminants such as PAHs, BTEX, chlorinated hydrocarbons (the last under anaerobic conditions), and other contaminants, despite that it ignores that transformation products may be hazardous too.

The importance of degradation can be appreciated from an example of groundwater (rather than soil) contamination. About, say, a decade ago, the concept of natural attenuation has been developed. This concept proposes that the subsoil environment is able to cause natural degradation of contaminants, e.g. due to the intrinsic activity of microbial populations. Although dispersional mixing and dilution, as well as volatilization of chemicals may contribute to natural attenuation, degradation is a major process in this concept. The concept as such is important as it diminishes the environmental hazards of soil contamination, and therefore has become a major issue in soil and groundwater contamination strategies, management, and decision making.

Full Simplified Mass Balance Equation

A full mass balance equation assuming linear equilibrium adsorption and first order decay can now be written as:

$$\frac{\partial}{\partial t}(nRC) + \nabla \cdot (qC) - \nabla \cdot (nD \cdot \nabla C) + n\lambda RC = 0 \tag{49}$$

where it has also been assumed that the component dissolved in the water phase as well as the component adsorbed onto the solid phase can both decay with the same decay constant. In reality, this is not necessarily true.

This equation is the foundation of most software aimed at modelling soil and groundwater contamination, such as MODFLOW/MT3D and related models. As such, this equation is the core of much scientific as well as management supporting investigations done at international, national and local levels. Few, if any, predictions and prognoses are made on the fate of contaminants, that are not based on equation (49).

For some one-dimensional and simple two- or three-dimensional problems, analytical solutions exist for equation (49).

With those solutions, it is easy to obtain an impression of the effects of the different processes and their parameters on the transport behaviour of dissolved chemicals.

Some Effects in the Numerical Solution of the Transport Equation

In many cases, analytical solutions of the simplified transport equation are not available. That is e.g. the case for heterogeneous systems or complicated boundary conditions. In those cases one has to resort to the numerical solution of the groundwater flow equation and the solute transport equation.

In order to analyse the behaviour of the numerical solution of the partial differential equation describing the transport of a solute, we consider a simplified 1-dimensional system with constant porosity n, constant specific discharge q and constant dispersion coefficient D:

$$\frac{\partial C}{\partial t} + v\frac{\partial C}{\partial x} - D\frac{\partial^2 C}{\partial x^2} = 0 \tag{50}$$

For the evaluation of the spatial derivatives in a finite difference approach, we use the following Taylor series expansion:

$$C(x + \Delta x) = C(x) + \Delta x\frac{\partial C}{\partial x} + \frac{(\Delta x)^2}{2}\frac{\partial^2 C}{\partial x^2} + \frac{(\Delta x)^3}{6}\frac{\partial^3 C}{\partial x^3} + \dots \tag{51}$$

$$C(x - \Delta x) = C(x) - \Delta x\frac{\partial C}{\partial x} + \frac{(\Delta x)^2}{2}\frac{\partial^2 C}{\partial x^2} + \frac{(\Delta x)^3}{6}\frac{\partial^3 C}{\partial x^3} + \dots$$

where Δx is the blocksize in the x-direction

The first order derivative of C with x can now be approximated in two ways. A backward finite difference approximation follows from the second equation given by (51):

$$\frac{\partial C}{\partial x} \approx \frac{C(x) - C(x - \Delta x)}{\Delta x} + \frac{\Delta x}{2}\frac{\partial^2 C}{\partial x^2} - \dots \tag{52}$$

Another approximation for the first order derivative can be obtained by taking the difference of the two equations given in (51):

$$\frac{\partial C}{\partial x} \approx \frac{C(x + \Delta x) - C(x - \Delta x)}{2\Delta x} + \frac{(\Delta x)^2}{6}\frac{\partial^3 C}{\partial x^3} + \dots \tag{53}$$

An approximation for the second order spatial derivative is obtained by adding the two equations (51). After some manipulation, the following approximation is obtained:

$$\frac{\partial^2 C}{\partial x^2} \approx \frac{C(x + \Delta x) - 2C(x) + C(x - \Delta x)}{(\Delta x)^2} + \frac{(\Delta x)^2}{24}\frac{\partial^4 C}{\partial x^4} + \dots \tag{54}$$

where a higher order term not given in equation (51) has been taken into account. The truncation error for the approximation of the first order derivative is for the backward difference (52) of the order Δx, and for the central difference of the order $(\Delta x)^2$. In other words, the central difference approximation is more accurate. The truncation error for the approximation of the second order derivative is of the order $(\Delta x)^2$.

For the time derivative the following approximation can be obtained:

$$\frac{\partial C}{\partial t} \approx \frac{C(t) - C(t - \Delta t)}{\Delta t} + \frac{\Delta t}{2}\frac{\partial^2 C}{\partial t^2} - \dots \tag{55}$$

where Δt is the time step and the derivative in the higher order term is evaluated at time $t+\Delta t$ (i.e. at the end of the time step). Note, that the truncation errors in the approximation for the spatial and temporal derivatives indicate, that small grid blocks (small Δx) should be used where the second order derivative of the concentration with respect to x is large, and small time steps should be used when the second order derivative with respect to t is large.

In the following, we will adopt the notation: $C_i=C(x)$, $C_{i-1}=C(x-\Delta x)$, $C_{i+1}=C(x+\Delta x)$, while values evaluated at the beginning of a time step will have a superscript o, and values evaluated at the end of a time step will have a superscript n.

If we evaluate the spatial derivatives at a time level between the beginning of the time step and the end of the time step, the discretised mass balance equation can then be written as:

$$\frac{C_i^n - C_i^o}{\Delta t} + v\left[\theta\frac{C_i^n - C_{i-1}^n}{\Delta x} + (1-\theta)\frac{C_i^o - C_{i-1}^o}{\Delta x}\right] -$$
$$D\left[\theta\frac{C_{i+1}^n - 2C_i^n + C_{i-1}^n}{(\Delta x)^2} + (1-\theta)\frac{C_{i+1}^o - 2C_i^o + C_{i-1}^o}{(\Delta x)^2}\right] = 0 \tag{56}$$

where θ is a factor between o and 1. For $\theta=1$, all spatial derivatives are evaluated at the new time level (the end of the time step). This is a fully implicit scheme. For $\theta=0$, all spatial derivatives are evaluated at the old time level (the beginning of a timestep). This is a fully explicit scheme. A mixed scheme (known as the Crank-Nicholsen scheme) is obtained by setting $\theta=0.5$.

Note that in equation (56) a backward difference for the advective term is used. A similar expression can be obtained for a central difference approximation for the advective term.

Numerical Dispersion

Numerical dispersion is an extra dispersion in the numerical solution of the transport equation which is caused by the discretisation of the advective term.

Consider the discretised equation (56), and evaluate all spatial derivatives at the end of the time step (fully implicit, $\theta=1$):

$$\frac{C_i^n - C_i^o}{\Delta t} + v\frac{C_i^n - C_{i-1}^n}{\Delta x} - D\frac{C_{i+1}^n - 2C_i^n + C_{i-1}^n}{(\Delta x)^2} = 0 \tag{57}$$

One could now ask the question which partial differential equation is approximately solved by these equations if higher order terms are taken into account. Using equations (52), (54) and (55), keeping all terms with derivatives of order 2 then gives the following partial differential equation:

$$\frac{\partial C}{\partial x} - \frac{\Delta t}{2}\frac{\partial^2 C}{\partial t^2} + v\frac{\partial C}{\partial x} - \frac{v\Delta x}{2}\frac{\partial^2 C}{\partial x^2} - D\frac{\partial^2 C}{\partial x^2} = 0 \tag{58}$$

In order to obtain an expression for the second order derivative with respect to time, we will differentiate equation (50) with respect to time:

$$\frac{\partial^2 C}{\partial t^2} = -v\frac{\partial^2 C}{\partial x \partial t} + D\frac{\partial^3 C}{\partial x^2 \partial t} \tag{59}$$

An expression for the cross derivative term is obtained by differentiating equation (50) with respect to x:

$$\frac{\partial^2 C}{\partial x \partial t} = -v\frac{\partial^2 C}{\partial x^2} + D\frac{\partial^3 C}{\partial x^3} \tag{60}$$

Substitution of equation (60) in (59) then gives:

$$\frac{\partial^2 C}{\partial t^2} = v^2\frac{\partial^2 C}{\partial x^2} - vD\frac{\partial^3 C}{\partial x^3} + D\frac{\partial^3 C}{\partial x^2 \partial t} \tag{61}$$

Substitution of (61) in (58), collecting the second order derivative terms and neglecting higher order terms then gives:

$$\frac{\partial C}{\partial t} + v\frac{\partial C}{\partial x} - \left[D = \frac{v\Delta x}{2} + \frac{v^2\Delta t}{2}\right]\frac{\partial^2 C}{\partial x^2} = 0 \tag{62}$$

Equation (62) is up to second order terms identical to the discretised equation (57). In other words, the discretised equation is an approximation to a solute transport equation with enhanced dispersion (cf. the term between brackets). Note, that in the analysis higher order terms are neglected. As a consequence, the discretised equation (57) is not completely identical to equation (62).

The extra dispersion $v\Delta x/2 + v^2\Delta t/2$ is called the numerical dispersion. If we assume that molecular diffusion can be neglected, the physical dipersion is given by $D = \alpha v$, where α is the dispersivity of the medium. The numerical dispersion can now be neglected if the block size Δx and the time step Δt are chosen such that:

$$\alpha \gg \frac{\Delta x}{2} + \frac{v\Delta t}{2} \tag{63}$$

Carrying out the same analysis with the advective term in equation (57) approximated by a central difference, will show that the numerical dispersion in that case is $v_2\Delta t/2$, which is smaller than the one for the backward difference of the advective term. There can be, however, reasons for adopting the backward difference approximation.

There are a number of ways in which we can neutralize the effect of numerical dispersion. The most obvious way is to correct the dispersivity for the numerical dispersion. Suppose the physical dispersivity is given by α_f, while the dispersivity defined for the numerical calculations is given by α_m. For a finite difference approximation with a backward difference for the advective term, the following choice:

$$\alpha_m = a_f - \frac{\Delta x}{2} - \frac{v \Delta t}{2} \tag{64}$$

will result in a total dispersion in the numerical calculations equal to the physical dispersion. This approach can, however, only be adopted if the model dispersivity α_m remains positive because negative values of the model dispersivity will generate instabilities in the numerical solution.

More complicated ways to minimise numerical dispersion in the numerical solution of the solute transport equation can be thought of. In all cases one should consider the fact that numerical dispersion is solely caused by the first order spatial derivative (the advective term).

One way of avoiding numerical dispersion is by what is called operator splitting. In that approach the change in concentration is split in two parts: one part due to advective transport, and one part due to dispersion:

$$\left(\frac{\partial C}{\partial t} \right)_a = -v \frac{\partial C}{\partial x}$$

$$\left(\frac{\partial C}{\partial t} \right)_d = D \frac{\partial^2 C}{\partial x^2} \tag{65}$$

and adding the two contribution gives the full transport. The first part in equation (65), the advective part is now solved by a characteristic method. In this method, water particles are followed as they are transported with velocity v. Each water particle represents a certain mass of solute, which in a time step Δt is transported over a distance $v \Delta t$. Once the advective transport has been solved, the dispersion has to be added. That again can be done in a number of ways, the most simple being a finite difference approximation. Another way is the random walk method, where the effect of dispersion is simulated by random displacements of the water particles around the mean displacement given by the velocity v. These random displacements are related to the dispersion coefficient.

Using a characteristic method to simulate the advective transport has the disadvantage that only a discrete number of particles can be followed, and that interpolation is required to transform mass per particle to concentration distribution. A smooth concentration distribution from the distribution of the particles can only be obtained if a very large number of particles is used. This is especially true for the relative low concentration contours.

A method that strongly resembles the characteristic method is the Eulerian-Lagrangian method. In this method, the time derivative is approximated by taking the difference in concentration not at the same place, but at different places. Using a Taylor series expansion, it can be shown that the expression:

$$\frac{C(x,t+\Delta t)-C(x-v\Delta t,t)}{\Delta t} \approx \frac{\partial C}{\partial t} + v\frac{\partial C}{\partial x} \tag{66}$$

is second order correct, i.e. neglected derivatives are of third order or higher. Consequently, no numerical dispersion is generated by this method. In a standard finite difference method the second term in the left side of (66) is evaluated by evaluating the concentration at position $x-v\Delta t$ at the beginning of a time step. That can usually be one by interpolation, although special precautions have to be taken close to boundaries.

Each of the methods mentioned here can easily be extended to two or three dimensions.

Oscillations in the Solution

Some of the finite difference schemes can, under certain conditions, generate oscillations in the solution, i.e. negative concentrations or concentrations larger than the maximum value defined by the initial and boundary conditions may occur. It should be pointed out that these are not instabilities, where errors in the solution can grow unbounded.

Consider the discretised, implicit equation, where the advective term is approximated by a backward difference (equation (57)). With some manipulation, this equation can be written as:

$$C_i^n = \beta_i C_i^o + \beta_{i-1}C_{i-1}^n + \beta_{i+1}C_{i+1}^n$$

$$\beta_i = \frac{1}{1+\dfrac{v\Delta t}{\Delta x}+\dfrac{2D\Delta t}{(\Delta x)^2}} \quad \beta_{i-1} = \frac{\dfrac{v\Delta t}{\Delta x}+\dfrac{D\Delta t}{(\Delta x)^2}}{1+\dfrac{v\Delta t}{\Delta x}+\dfrac{2D\Delta t}{(\Delta x)^2}} \tag{67}$$

$$\beta_{i+1} = \frac{\dfrac{D\Delta t}{(\Delta x)^2}}{1+\dfrac{v\Delta t}{\Delta x}+\dfrac{2D\Delta t}{(\Delta x)^2}}$$

Equation (67) shows that the concentration at the new time level is a weighted average of the concentration at the old time level and the concentrations in the adjacent grid blocks. Inspection of the weighing coefficients β shows that:

$$\beta_i + \beta_{i-1} + \beta_{i+1} = 1 \quad \text{and} \quad \beta_i, \beta_{i-1}, \beta_{i+1} > 0 \tag{68}$$

As a consequence C obeys a maximum principle, i.e. it can never become smaller that the smallest value given in the initial and boundary conditions, or larger than the largest value given in the initial and boundary conditions.

If we would have used a central difference for the advective term an expression similar to equation (67) can be written, however, with weighting coefficients:

$$\beta_i = \frac{1}{1 + \dfrac{2D\Delta t}{(\Delta x)^2}}$$

$$\beta_{i-1} = \frac{\dfrac{D\Delta t}{(\Delta x)^2} + \dfrac{v\Delta t}{2\Delta x}}{1 + \dfrac{2D\Delta t}{(\Delta x)^2}} \tag{69}$$

$$\beta_{i+1} = \frac{\dfrac{D\Delta t}{(\Delta x)^2} - \dfrac{v\Delta t}{2\Delta x}}{1 + \dfrac{2D\Delta t}{(\Delta x)^2}}$$

Inspection now reveals that the sum of the weighting factors β again equals 1, but that all weighting factors are positive only under the condition:

$$\frac{D\Delta t}{(\Delta x)^2} > \frac{v\Delta t}{2\Delta x} \quad \text{or} \quad D > \frac{v\Delta t}{2} \quad \text{or} \quad \alpha > \frac{\Delta x}{2} \tag{70}$$

Condition (70) is often given in a slightly different form:

$$Pe_{cell} = \frac{v\Delta x}{D} < 2 \tag{71}$$

where Pe_{cell} is called the cell Peclet number (equation (37)).

Basically, a backward difference for the advective term generates more numerical dispersion than a central difference. However, a central difference approximation might result in oscillations in the solution, which will not occur for a backward difference.

It should be pointed out that the global mass balance will in all cases by perfect, irrespective of numerical dispersion or oscillations. This also indicates that damping oscillations in a solution by simply not allowing the concentration to become larger than a predefined value or smaller than another predefined value will ultimately result in mass balance errors.

Stability of the Explicit Solution

The explicit formulation of the discretised equations has the advantage that an explicit expression for the concentrations in each grid block is obtained, which therefore does not require matrix manipulation to obtain the solution. However, under certain conditions, such an explicit formulation may become unstable, i.e. small errors in the solution may grow uncontrolled and unbounded in time, resulting in very large positive and negative concen-trations.

Appendix A gives the derivation of the criteria for the explicit formulation to be stable. In general, conditions $\beta_2 < 1$ or $\Delta t < \dfrac{(\Delta x)^2}{2D}$ and $\beta_2 > 4\beta_1^2$ or $\Delta t < \dfrac{2D}{v^2}$ are given in a slightly different form. Condition $\beta_2 < 1$ or $\Delta t < \dfrac{(\Delta x)^2}{2D}$ can be written as:

$$2D < \frac{(\Delta x)^2}{\Delta t} \tag{72}$$

Substitution of this relation in condition $\beta_2 > 4\beta_1^2$ or $\Delta t < \dfrac{2D}{v^2}$ then gives:

$$\Delta t < \frac{2D}{v^2} < \frac{(\Delta x)^2}{v^2 \Delta t} \quad or \quad \frac{v^2(\Delta t)^2}{(\Delta x)^2} < 1 \tag{73}$$

which can then be given as the well-known Courant condition:

$$\frac{v\Delta t}{\Delta x} < 1 \tag{74}$$

and the Neumann condition:

$$\frac{D\Delta t}{(\Delta x)^2} > 0.5 \tag{75}$$

Both conditions have a physical interpretation. The Courant condition states that in one time step, a water particle cannot travel further than the length of a grid block. The Neumann condition relates the characteristic length associated with the dispersion over a time step to the block size.

For a stable solution it is required that both conditions are satisfied, which means that the most restrictive condition determines the time step size that will still result in a stable solution.

Even an instable solution will give a perfect global mass balance (provided we are able to calculate the mass balance with enough significant digits). Consequently, a perfect mass balance (very small errors) is no guarantee for a good solution. However, large errors in the global mass balance is a guarantee for errors in the concentration distribution.

Initial and Boundary Conditions

In order to be able to model the transport of (reactive) solutes in groundwater it is necessary to define both the initial and boundary conditions. Initial conditions are (mathematically) only required for transient or time-dependent problems.

For local pollution problems, the initial conditions (a clean soil) can usually be given for the time before the pollution or spill occurred. For diffuse sources of pollution, the initial condition (present day situation in cases predictions have to be made) are derived by simulating long periods before the present day, using known (or estimated) mass inflow.

Boundary conditions for solute transport can be defined similar to the boundary conditions for groundwater flow. We distinguish three types of boundary conditions:

- Dirichlet: the concentration on the boundary is fixed. Although mathematically this is a valid boundary condition, it is physically almost always impossible to create such a boundary condition. Nevertheless it is often applied at inflow boundaries, where the concentration of the solute in the water phase is known.

- Neumann: the total mass flow of a solute across a boundary is defined. This type of boundary condition (also called mass loading) is often applied to inflow boundaries in e.g. experiments, but also in field situations dealing with point sources of pollution where the total mass of solute entering the groundwater is known (or can be estimated). For outflow boundaries, this type of boundary condition is physically not possible, because at outflow boundaries the total mass leaving the system is unknown (dependent on the concentration in the aquifer).

- Cauchy: the mass flux of solute across the boundary is dependent on the concentration in the water phase itself. This type of boundary condition is often applied at outflow boundaries, defining the total mass flux as qC, where q is the water flux and C is the concentration of the solute in the water. Implicit in this definition of the boundary condition is the assumption that the dispersive flux across the boundary is zero.

For local point source pollution problems, we choose boundaries far enough away from the point source to make sure that the boundary conditions do not influence the concentration distribution.

Non-linear, Non-equilibrium Processes

General

Diffusion and radio-active decay have already been described in section 2, and do not need further elaboration.

Advection and Dispersion

In heterogeneous systems, advection and dispersion are closely related. Dispersion is used to describe the transport of a contaminant due to variations in the groundwater flow which are not described by our "model" (where model can mean anything from complex numerical systems to the assumption of uniform flow). These variations are caused by local heterogeneities which have not been taken into account

Basically, this means that the more information on convective or advective transport (in fact the variation in hydraulic conductivity) are taken into account, the smaller the dispersion will be.

From large numbers of field measurements, it is known that within one geological formation, the hydraulic conductivity may show a log-normal distribution with for instance an exponential covariance function defining the spatial correlation:

$$Cov(Y_1, Y_2) = \sigma_Y^2 e^{-\frac{r}{l_Y}} \tag{76}$$

where $Y=ln(k)$, σ_Y is the standard deviation in Y, r is the distance between the points where Y_1 and Y_2 were measured, and l_Y is the correlation length. The basic assumption is that the covariance is dependent on the distance between the measurement points only. This relation can be extended to account for direction dependence. For instance, the fact that sedimentation in geological formation have taken place in a certain direction will generate this direction dependence (correlation length in the z-direction will be smaller than the correlation length in the x and y-directions).

If we consider the formation to be homogeneous, all heterogeneities in the formation will have to be described by dispersion. In such a system, the dispersivity will be related to the correlation length (relation dependent on the flow pattern). That is, however, only true if the plume "has seen" all heterogeneities, i.e. if the size of the plume is (much) larger than the correlation length. In that case we will call the plume "ergodic". For small plumes (non-ergodic), that is not the case, and the behaviour of such a plume cannot be described on a large scale by global dispersion. For the behaviour of such plumes it is necessary to incorporate information on local heterogeneities.

Note, that hydrodynamic dispersion generates a spreading in the average values of the concentrations, where the averaging volume is determined by the scale on which we assume the system to be homogeneous. If that scale is large, we are in principle not allowed to make a comparison of calculated concentrations with local measurements. Another reason to use spatial moments.

Adsorption/Desorption

Equilibrium adsorption/desorption, especially for larger concentrations, can often be described by either a Freundlich equation

$$C_s = K(C)^p \tag{77}$$

where K is a constant, and $p<1$, or by a Langmuir isotherm:

$$C_s = \frac{C_{s\,max}kC}{1+kC} \tag{78}$$

where $C_{s\,max}$ the maximum amount is that can be adsorbed (occurs when $kC>>1$). For low concentrations (where $kC<1$) relation (78) becomes linear. The Langmuir isotherm is typical for soils and solid surfaces that have a limited number of sites available for adsorption.

For the non-linear adsorption, we can still define a retardation factor:

$$R = 1 + \frac{1-n}{n}\rho_s\frac{dC_s}{dC} \tag{79}$$

which will, however, be dependent on the concentration C. The mass balance equation is then given by:

$$R\frac{\partial C}{\partial t} + \nabla \cdot F = 0 \tag{80}$$

where F is the mass flux (by advection and dispersion).

Inspection of the Freundlich and Langmuir isotherms shows that the retardation factor becomes smaller for larger concentrations. For the displacement of a front of solute, this means that the higher concentrations are less retarded than the lower concentrations. In other words, the higher concentrations try to "overtake" the lower concentrations. This effect is counteracted by dispersion. After some time, equilibrium will occur between these competing processes and a travelling wave will develop (Bosma and Van der Zee). An example of such displacement is shown in figure below.

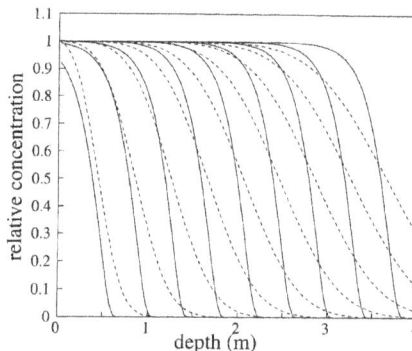

Traveling wave type of displacement, with concentration fronts given by solid lines, compared with linear convective-dispersive transport, with fronts given by dashed lines

In case of non-equilibrium, the adsorption/desorption may also be non-linear due to the limited number of sites available. For non-equilibrium adsorption/desorption we also need to solve for a mass balance of the adsorbed solute. For instance, a Langmuir type non-equilibrium interaction takes the form:

$$I = k_a (1 - \theta)C - k_d C_s \tag{81}$$

where θ is the fraction of the adsorption sites that is occupied. Non-equilibrium often results in tailing in breakthrough curves.

If more solutes are present, competition for the adsorption sites may occur. One of the effects that may occur, is, that some solute adsorbs fast, but will later be replaced by a solute that adsorbs slower, but has a higher affinity for the adsorption sites.

Chemical Reactions

Chemical reactions are usually described by equilibrium reactions. These are in general highly non-linear, while a large number of species play a role. However, they are local in nature (there is no spatial partial derivative in these equations). Solving for chemical reactions in combination with transport can be done in two ways:

1. Operator splitting: solve for the transport of all species required. The result is a redistribution of the concentrations. Starting with these concentrations, calculate the new chemical equilibrium, etc. Advantage is that the number of equations is limited. Disadvantage is that the time step size is limited.

2. Combine the equations for chemical equilibrium with the transport equations, and solve simultaneously. Advantage is that the time step size is less limited than in the operator splitting method. Disadvantage is the large system of non-linear equations that need to be solved.

Note that negative concentrations (however small they are) cannot be allowed when dealing with chemical reactions.

Dissolution/precipitation reactions play a slightly different role. These reactions are dependent on threshold values. Consider a mineral the dissolves in the water phase. As long as the mineral is present, the concentration in the water phase is constant, and the mass transfer from the solid to the liquid phase is unknown. When the mineral is not present, it will act in the water phase as any other species. Until the concentration becomes large, and precipitation starts to occur.

One of the biggest problems in chemical reactions is to limit the number of species (and hence the number of mass balances that need to be considered) that have to be taken into account.

Biodegradation

Biodegradation needs at least, beside the present of bacteria, also the presence of a carbon source and the presence of an electron acceptor (if degradation is aerobic). The degradation can also be anaerobic (e.g. of inflammable NAPLs such as chlorinated hydrocarbons), in which a chemical is needed that supplies electrons for the degradation process. In general, aerobic degradation is faster than anaerobic degradation.

- Biodegradation can be assumed to take place in the water phase only. Dependent on the number of bacteria present, we can distinguish between:

- single bacteria in the water phase; these can also be transported with the flowing ground-water;

- colonies on the solid matrix; these bacteria are not transported, but they still have access to all species in the water phase;

- biofilms on the solid matrix; for bacteria in biofilms, mass transport from the free water phase to the biofilm is usually diffusion controlled, and this (slow) mass transfer has to be taken into account when describing biodegradation.

If a carbon source is available, the biomass (number of bacteria) will grow, and in time we can have a transition from free bacteria to colonies to biofilm. Two other effects play a role in biodegradation: 1) bacteria die, which can usually be described by a first order decay, and 2) species may be present that inhibit the biodegradation.

If a first order description of biodegradation is not sufficient, a typical way to describe biodegradation then is e.g. by Monod type relations:

$$I = I_{max} \left(\frac{C}{k_C + C} \right) \left(\frac{O}{k_O + O} \right) \left(1 - \frac{I}{k_I + I} \right) \tag{82}$$

where I_{max} is the maximum amount that can be degraded, C is the concentration of the carbon source, O the concentration of the electron acceptor, I the concentration of the inhibitor, and k are constants. Equation (82) can be extended to include more species that play a role in the biodegradation. Note, that for each of these species a mass balance needs to be solved. Equation (82) does not include the biomass, but that can be done in the same way, provided a mass balance for the biomass is solved as well.

If biodegradation takes place, the products of the degradation may in itself be degraded again by the same (type of) bacteria. In these cases it is required to analyse (or predict) the transport of multiple solutes, where the degradation of one solute results in a source term for the daughter product. These multiple solutes will in general have different adsorption/desorption characteristics and different decay characteristics.

Non Aqueous Phase Liquids (NAPL)

Non-aqueous phase liquids or oil does not mix with water. Infiltration of these liquids in the subsurface (e.g. through leaking tanks or pipes) will create multiphase systems.

If a spill occurs, oil will be transported through the unsaturated zone, leaving behind a residual oil saturation (around 20-30% of the pore volume), which is not mobile due to capillary forces.

When the oil reaches the water table the oil will float on the water table if it is lighter than water (LNAPL). Such floating lenses have been found on many places worldwide, often due to spills of gasoline at gasoline stations. The analysis of such lenses has been studied experimentally (Wipfler et al., 2004) but also numerically and analytically (Van Dijke and Van der Zee, 1998). If the NAPL is denser than water (DNAPL), it will move downwards through the groundwater until it reaches an impermeable layer. The transport of DNAPL through the saturated zone is controlled by instabilities (local heterogeneities), and it is therefore very difficult to predict where exactly the DNAPL will be present. However, residual oil will be present at those locations where the DNAPL has passed through.

Components of the oil will dissolve in the water phase, although generally in very small quantities. If the oil phase is in equilibrium with the water phase, the concentration of the oil components in the water phase can be described by Raoult's law:

$$C = C_{max} X \tag{83}$$

where C_{max} the maximum concentration is if the water phase is in contact with the pure component, and X is the mol fraction of the component in the oil phase.

For a layer of the pure oil, there is in general no equilibrium between the water phase and the oil phase. In these cases, the mass transfer is diffusion controlled, and described by a first order relation:

$$I = k(C - C_{eq}) \tag{84}$$

where k is the mass transfer coefficient, and C_{eq} is the equilibrium concentration, which is given by equation (83). The mass transfer coefficient is dependent on a.o. the diffusion coefficient, the oil saturation, average pore diameter and the groundwater velocity. A number of empirical relations exist that define these relations.

If a free oil phase is present, the transport of oil components requires both the solution of the oil and water flow equations and the transport equation of each species in both phases. Note, that the oil and water flow are coupled due to the capillary forces, and the dependence of the water hydraulic conductivity on the oil saturation.

Density Dependent Flow and Transport

In many cases, the properties of groundwater, and in particular the density, are influenced by the concentration of dissolved species. That is e.g. the case for sea water intrusion in coastal aquifers, or for the infiltration of leachate from a landfill into a fresh water aquifer. In many of these cases, we may assume that the density is the only property of the water phase that is influenced by the concentrations of the dissolved species. In other cases, we have to take into account the fact that the viscosity of the water phase is also influenced by the concentration of the dissolved species. That is in particular true for very high solute concentrations (or for large temperature differences). For instance, if we consider the disposal of waste in deep saline aquifers, the effect of a changing viscosity cannot be neglected. The viscosity of water varies by a factor of 2 from fresh water to concentrated brine. These situations differ from those of NAPLs, as the different types of groundwater are miscible, whereas NAPL and water are immiscible.

In case of changing fluid properties, the governing flow equation cannot be formulated in terms of groundwater potential but have to be given in terms of pressures and a gravity term. In the following, the governing equations for density dependent flow and transport will be given, followed by a simplified set of equations.

Basic Equations

For density dependent flow, the general form of Darcy's law needs to be used:

$$q_x = -\frac{\kappa_x}{\mu}\frac{\partial p}{\partial x} \quad q_y = -\frac{\kappa_y}{\mu}\frac{\partial p}{\partial y} \quad q_z = \frac{\kappa_z}{\mu}\left(\frac{\partial p}{\partial z} + \rho g\right) \tag{85}$$

The mass balance equation for the water phase is given by:

$$\frac{\partial}{\partial}(n\rho) + \nabla \cdot (\rho q) = 0 \tag{86}$$

and the mass balance for the solute by:

$$\frac{\partial}{\partial t}(n\rho\omega) + \nabla \cdot (\rho\omega q) - \nabla \cdot (n\rho D \cdot \nabla\omega) = 0 \tag{87}$$

where it has been assumed that there are no external sources and sinks, and that the solute creating the density differences is a conservative one.

We can now define a "fresh water potential" as:

$$h_f = \frac{p}{\rho_f g} + z \tag{88}$$

where ρ_f is the density of fresh water ($\omega = 0$) and z is the vertical position with respect to a reference level. Substitution of (88) in (85) gives the following form of Darcy's law:

$$q_x = -\frac{\kappa_x \rho_f g}{\mu}\frac{\partial h_f}{\partial x} \quad q_y = -\frac{\kappa_y \rho_f g}{\mu}\frac{\partial h_f}{\partial y} \quad q_z = -\frac{\kappa_x \rho_f g}{\mu}\left(\frac{\partial h_f}{\partial z} + \frac{\rho - \rho_f}{\rho_f}\right) \tag{89}$$

which shows that the hydraulic conductivities are dependent on the fluid properties. It is also obvious that the fresh groundwater potential is not the only driving force for the groundwater flow. The system of equations (85) or (89) with (86) and (87) have to be supplemented by equations of state for both the density and the viscosity of the water.

The mass balance equations with Darcy's law form a set of coupled, non-linear equations. The coupling is a two-way coupling: the flow is dependent on the concentration distribution (through the dependence of the density and the viscosity on the concentration), while the concentration distribution is dependent on the flow through the advective term in equation (87).

Simplified Equations

In many cases, the equations governing the density dependent flow and transport can be simplified by making a number of assumptions. First of all, for relatively low concentrations (e.g. the salt concentration in seawater) we may assume that the changes in the viscosity with the concentration are negligible. Furthermore, the density can be assumed to be linear dependent on the salt mass fraction. That is certainly the case for salt water:

$$\rho = \rho_f \left(1 + \gamma \omega\right) \tag{90}$$

where for salt, γ has a value of 0.7.

Secondly, we will assume the (temporal) changes in the porosity can be neglected.

Finally, we can adopt Boussinesq's approximation, which states that the variations in the liquid density can be neglected everywhere, with the exception in the gravity term of Darcy's law. With all these assumptions, the governing equations can be written as:

Mass balance of the water phase:

$$\nabla \cdot q = 0 \tag{91}$$

Mass balance of the salt:

$$\frac{\partial(n\omega)}{\partial t} + \nabla \cdot (\omega q) - \nabla \cdot (nD \cdot \nabla \omega) = 0 \tag{92}$$

Darcy's law:

$$q = -\frac{\kappa}{\mu}(\nabla p + \rho g e_z) \tag{93}$$

Equation of state:

$$\rho = \rho_f (1 + \gamma\omega) \tag{94}$$

where e_z is a unit vector in the z-direction (positive upward).

Substitution of the equation of state (94) in Darcy's law (93), and assuming that the porosity and the dispersion coefficients are constants, even further simplifies the set of equations:

$$\nabla \cdot q = 0 \tag{95}$$

$$n\frac{\partial(\omega)}{\partial t} + \nabla \cdot (\omega q) - \nabla \cdot nD\nabla^2\omega = 0 \tag{96}$$

$$q = -\frac{\kappa}{\mu}(\nabla(p + \rho_f gz) + \rho_f \gamma\omega g e_z) \tag{97}$$

These equations can be made dimensionless by choosing appropriate reference values for the different variables in the equations:

$$q_d = \frac{q}{q_r} \quad x_d = \frac{x}{L} \quad t_d = \frac{t}{t_r} \quad \omega_d = \frac{\omega}{\omega_r} \quad p_d = \frac{p + \rho_f gz}{p_r} \tag{98}$$

where L is a characteristic dimension of the system considered.

Now, if we choose the following reference values:

$$q_r = \frac{nD}{L} \quad t_r = \frac{L^2}{D} \quad \omega_r = \omega_{max} \quad p_r = \frac{\mu L^2}{\kappa nD} \tag{99}$$

the system of equations reduces to:

$$\nabla \cdot q = 0 \tag{100}$$

$$\frac{\partial\omega}{\partial t} + \nabla \cdot (\omega q) - \nabla^2 \omega = 0 \tag{101}$$

$$q = -\nabla p - A\omega e_z \tag{102}$$

where it is understood that all variables are dimensionless (subscript d has been omitted), and:

$$A = \frac{n\kappa D\Delta\rho_{max}}{\mu L} \tag{103}$$

is the Rayleigh number. This number defines the ratio of gravity and dispersive forces. Together with the initial and boundary conditions it fully controls the solution of equations (100) through (102).

For the stable situation, where fresh water (derived from rainfall) is situated above saline groundwater, analyses for the behaviour of such rainfall lenses in coastal delta areas.

For systems that are possibly instable, i.e. systems where fresh water is overlain by water with a higher density, the Rayleigh number controls whether such instabilities will occur. These situations arise e.g. under landfills, where the (heavier) leachate infiltrates in a fresh water aquifer, or at transgression of the coast line by rise of the seawater level.

For reasonable simple geometries, a perturbation analysis can show us when instabilities may occur. For the Rayleigh-Bernard problem, i.e. a rectangular vertical slab, with no-flow boundary conditions at the sides, and $\omega=1$ at the top boundary and $\omega=0$ at the bottom boundary, it can be shown that instabilities will occur for $A>4\pi^2$

Transport in Heterogeneous Media

If we wish to convey our understanding of transport of solutes to real soils, we have to account for spatiotemporal variability. After all, it is well known that soils vary spatially (layers, horizons, pores size distribution), and as a function of time, e.g. due to time varying weather, groundwater flow and many other causes. Such variability leads to rather complex behaviour in space and time. Instead of the rather simple, smooth concentration profiles and break-through curves, that we obtain for homogeneous soil and simple initial and boundary conditions, quite involved transport trajectories and concentration distributions result.

These are difficult to convey to the stakeholders, that have to base decisions on such results. For such reasons, spatial and temporal moment theory can be used to capture the transport phenomena in relatively robust terms. The moment theory is explained and for the case of spatial moments, it is presented mathematically. This is not done for temporal moments, because the mathematical details are completely in analogy to the spatial moments, and therefore obsolete here. Before going into details, first a qualitative impression is given of the moments.

The first spatial moment is related to the solute velocity, or, in case of a conservative solute, the groundwater velocity. For an instantaneous release of a non-reactive tracer in a steady state groundwater flow field, the first spatial moments as a function of time tell us exactly where the tracer is located and therefore also what the groundwater velocity is. Comparison of the first spatial moment of a reactive solute with the first spatial moment of a non-reactive tracer gives an estimate of the retardation factor. For a solute plume, these first moments characterize therefore the velocity of the entire plume, not of individual solute particles.

Following the zero'th spatial moment of a solute in time gives information about the degradation of the reactive solute. It is even better to compare these moments with the zero'th moment of a non-reactive (inert) tracer. If we indicate the properties of the reactive tracer with superscript r and the properties of the non-reactive tracer with n, we can define:

$$F^r = \frac{M_o^r(t)}{M_o^r(o)}$$

$$F^n = \frac{M_o^n(t)}{M_o^n(o)} \tag{104}$$

The decay constant λ of the reactive solute is then given by:

$$\frac{F^r}{F^n} = e^{-\lambda t} \tag{105}$$

We could have used the zero'th spatial moment of the reactive solute only (as a function of time). However, comparing it with the zero'th spatial moment of a non-reactive solute may give a better answer because possible errors due to limited available information may cancel if the concentrations required for the determination of the zero'th spatial moment are measured at the same locations.

The second central spatial moment is related to the dispersion on the scale of the plume. For an instantaneous release of a solute in a steady state groundwater flow field, one of the elements of the dispersion coefficient is given by:

$$D_{xx} = \frac{1}{2}\frac{d}{dt}\left(\sigma_{xx}^2\right) \tag{106}$$

and similar expressions can be given for the other 5 elements of the dispersion tensor (cf equation (33)).

The accuracy with which the spatial moments can be determined is of course dependent on the amount of data available. Higher order spatial moments are often less accurate than the lower order spatial moments.

Characterizing the behaviour of a plume of contaminant on the basis of spatial moments is usually more robust than trying to characterize it by individual measurements of concentrations. The latter are very dependent on local heterogeneities. Apart from that, decision makers are usually interested in global measures like the transport of the plume (first spatial moment) and the spread around the mean travel distance (second central spatial moment) rather than in local concentrations.

Moment Theory

As flow and transport conditions are spatiotemporally variable, complex patterns can develop, as illustrated in figure below for the distribution of a chemical that moves through soil. This pattern is difficult to describe to someone who cannot see it. However, often that is not really important, because we are only interested in simpler information. Simpler information, that is also statistically robust, are the so-called moments. Moments can be described for any property that is distributed (in space, in time). For instance, if we consider the concentration of a solute as in Figure below, that is distributed spatially, we can regard this distribution as a probability density function pdf (or: frequency function), in particular: a pdf of travelled distances, in this case. Its discrete analogue is a histogram, provided the lengths of all bars are normalized in such a way that their sum is equal to 1. A pdf f is related with a cumulative distribution (P) according to:

$$P = \int_{-\infty}^{x} f(\xi)d\xi \tag{107}$$

and for integration to $x = +\infty$, P becomes one: it is the probability that the value of x lies between the lower and the upper boundary, and this probability is one for this $x = +\infty$ upper boundary. Therefore, if we deal with a concentration distribution. We do so with the zero'th moment.

Instead of doing so, we will illustrate moments on mass instead of concentration. A concentration is a mass (of solute) divided by a mass or volume of water. You can of course calculate what happens if you add one glass of water with a particular concentration to another glass with another concentration. However, physically it is much better to go to more basic properties: how much mass of solute is in each glass, how much water, and what is the final mass of solute and of water if both are combined? Hence: never calculate moments of concentrations, or water fractions. Instead, calculate moments of quantities (kg, volume,.). For the present purpose, we calculate moments of solute mass in solution ($\theta(x, y, z, t).C(x, y, z, t)$) instead of concentration.

The zero'th moment of the solute mass distribution in solution is given by:

$$M_0(t) = \int_{-\infty}^{\infty} \theta(x,t)C(x,t)dx \tag{108}$$

if water fraction and concentration are only distributed in the x-direction (otherwise, we have to integrate for x, y, and z). The zero'th moment is also known as the mass of the distribution, and is a 'normalizing' property: division of higher order moments by the mass, renders the distribution of water fraction times concentration.

Evidence of spatial differences in the rate of propagation of a solute front. Part of a photograph kindly provided by R.G. (Gary) Kachanoski (Canada) of a loamy soil profile in Ontario, Canada. Front depth (white arrows) is visible because the lime rich subsoil is lighter coloured than the topsoil, where lime has dissolved and leached due to natural and anthropogenic acid inputs during the Holocene. Topsoil pH about 6.5, subsoil pH about 8.2.

The first moment is given by:

$$M_1(t) = \int_{-\infty}^{\infty} x \theta(x,t) C(x,t) dx = \int_{-\infty}^{\infty} x f_{\theta C}(x,t) dx \tag{109}$$

and written this way, illustrates that division by the 0th moment is needed to obtain a pdf. The first spatial moment divided by the oth spatial moment is known as the average position of the plume of solute (x_{av}), whereas the first temporal moment divided by the 0th temporal moment is the mean breakthrough time. The second moment is obtained by multiplying with x^2 instead of x. More common to use is the second central moment, given by:

$$M_2^c(t) = \int_{-\infty}^{\infty} (x - x_{av})^2 \theta(x,t) C(x,t) dx = \int_{-\infty}^{\infty} (x - x_{av})^2 f_{\theta C}(x,t) dx \tag{110}$$

from which the variance can be obtained as:

$$\sigma_{xx}^2(t) = \frac{M_2^c(t)}{M_0(t)} \tag{111}$$

This variance is a good measure for the spreading in space (here the x-direction), and is related with the concept of dispersion, as mentioned before. To describe the transport behaviour of solutes, the moments up to the variance are not always sufficient. The main reason is that the transport velocities, i.e., including retardation effects, of parcels of solute may not be symmetrically (e.g. normally) distributed. In that case, higher order moments may be needed, or the assumption that the pdf has another than normal shape is required. An important alternative to the normal (symmetric) pdf is the lognormal pdf.

The Lognormal Pdf

The normal pdf of X is given by:

$$f(X) = \frac{1}{s_X \sqrt{2\pi}} e^{\left(\frac{X - m_X}{s_X \sqrt{2}}\right)^2}$$

(112)

where s_X is the standard deviation and m_X is the mean. The lognormal pdf of Y is very similar, namely:

$$f(Y) = \frac{1}{Y \, s_Y \sqrt{2\pi}} e^{\left(\frac{Y - m_Y}{\sigma \sqrt{2}}\right)^2}$$

(113)

where the mean m_Y and standard deviation s_Y are those of the lognormally transformed parameter $X = \ln(Y)$. That means that if Y is lognormally distributed, then X is normally distributed, and the pdf of lognormally distributed Y is characterized with the statistics of X. The following equation shows that this is correct:

$$f_Y \, dY = \frac{1}{Y} f_{lnY} dY = f_{lnY} d(\ln y) = f_x \, dX$$

(114)

So, if we wish to characterize the lognormal pdf, we first have to log-transform, to calculate the statistics, and then we know the pdf. There is another way, as the statistics of both pfd's are related. The following relationships can be derived if Y is lognormally distributed and if $X = \ln(Y)$

$$m_X = \ln(m_Y) - \frac{1}{2} s_X^2$$

(115)

$$s_X^2 = \ln\left(1 + \left(\frac{s_Y^2}{m_Y^2}\right)\right)$$

So, to calculate the mean, you first have to calculate the variance (of X), from the statistics (mean and variance) of Y.

Median and mode for the normally distributed X are equal to the mean.

If you know the statistics of X (for instance log(hydraulic conductivity)), but not those of hydraulic conductivity itself (Y), then you can calculate those easily by inverting the above equations. The result is given by:

$$M_Y = e^{\left(m_X + \frac{1}{2} s_X^2\right)}$$

(116)

$$s_Y^2 = e^{\left(2m_X + \frac{1}{2}s_X^2\right)}\left(e^{s_X^2} - 1\right)$$

and for completeness, also the median and modus:

$$median_Y = e^{m_X}$$

$$mode_Y = e^{\left(m_X - s_X^2\right)} \qquad\qquad (117)$$

The lognormal pdf is very attractive for several reasons. An important reason is that many properties appear to be well lognormally distributed, whereas they are not normally distributed. For instance, the saturated hydraulic conductivity K_s has been found experimentally to be often lognormally distributed, and Miller and Miller also give a theoretical basis with the similar media or similitude theory.

Approaches to Model Transport in Heterogeneous Media

Modeling of heterogeneous media can be done in several ways, that vary in complexity. In all cases, an impression of the heterogeneity is gained experimentally. This gives us information regarding the statistics of the important soil properties, such as hydraulic conductivity. This information can be directly used as input for any model. However, it is also possible to generate a field of hydraulic conductivities and use that for input. This is attractive, because often we have data only for a very limited number of positions, leaving large volumes for which we do not have data and for which we have to find a good strategy to parameterize.

If we generate random fields, we can do a calculation for each generated field. If we do so repeatedly, for a designated set of statistics, then the calculations will be similar but different: just as when you take a picture of a meadow every five minutes: the meadow will be the same, but each picture is different. Repeated calculations for the same statistics is what we call Monte Carlo simulation. With this technique, we can determine the uncertainty that is intrinsic to our model results.

Monte Carlo simulation is a numerical approach, with which we can check whether analytical solutions are sound. But Monte Carlo simulation may also be done using analytical solutions, if they are available. Besides Monte Carlo, a common way to deal with heterogeneity is in the context of a GIS system, where each cell or pixel gets its own value, depending on overlay maps and transfer functions. Then, for each cell or pixel the same set of calculations is done, only with different parameters. An example of the output obtained this way is shown in figure for pesticide leaching. In these calculations, geospatial data of soil type, weather conditions, geohydrology, land use (which pesticides, which crops, which growing season and pesticide application date), and other information can be linked. Clearly, then it is possible to determine which regions have a major hazard of leaching a particular pesticide (or more generally, a contaminant of interest).

This type of modelling is very important for risk analysis and screening of pesticides in the admission of these chemicals. Whether or not a pesticide is admitted to the market, or should be taken out of the market, may depend on the outcome of fate calculations with models as described here. For instance, the pesticide screening in EU uses the so-called FOCUS protocols.

In a relatively simple approach, the spatial configuration is neglected. This is suitable, if the interest is not where something happens but what its effect is on the entire system. An example is leaching of solute from a field: instead of wishing to know where the leaching occurs, it may be sufficient to know what the average leaching is. This is the essence of the parallel stream tube PST model.

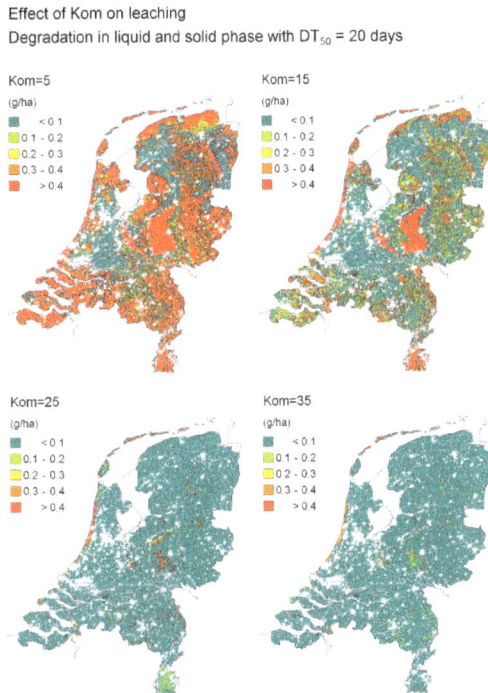

Effect of Kom on leaching
Degradation in liquid and solid phase with DT_{50} = 20 days

Leached pesticide into groundwater, as calculated with the EU pesticide screening model GEO-PEARL, for four Kom-values and a half-life of 20 days. Calculations and figure kindly provided by A. Tiktak (PBL, Bilthoven, Nether- lands). The assumption has been made that only dissolved pesticide may degrade

Parallel Stream Tube Model

This approach has been developed by Bresler and Dagan and their analyses for a conservative tracer was later extended to reacting chemicals. It is very effective to capture the effect that fronts do not move with the same velocity for all places. It is used more for leaching in unsaturated soil, than for groundwater transport, because stream tubes are more unidirectional in soil (vertical) than in groundwater.

For the vertical transport of solute, the simplest model is the purely convective model. For the case of linear sorption, it is given by:

$$z_c = \frac{vt}{R} \tag{118}$$

for each concentration that at time 0 was located at depth (z) equal to zero. For gravity leaching, the velocity is determined by the unit gradient and the hydraulic conductivity, where the latter has been demonstrated to be strongly spatially variable. The retardation factor is likewise known to vary spatially. This implies, that even if all concentrations 'start' at time 0 at the same position (say z=0), then after a certain time, they will have moved more in some stream tubes than in others. This is illustrated in figure. Due to spatial variability of flow velocity and retardation factor, also the front moves with a distributed velocity.

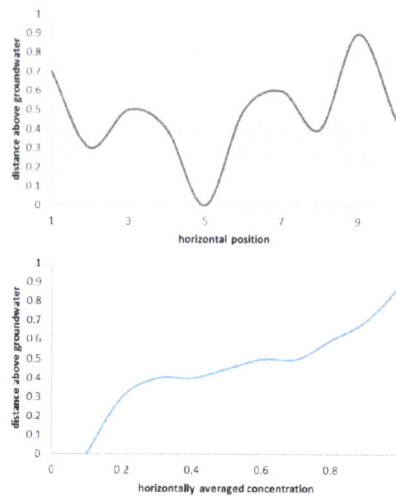

Illustration of the PST model for unsaturated soil, and Heaviside input of solute. The top panel was obtained by calculating the front position for many horizontal positions, giving a front position with above a relative concentration of 1 and below of 0. The bottom panel was obtained by calculating the fraction of the horizontal plane, for which the relative concentration is 1. This fraction is then the horizontally averaged (relative) concentration.

Now, it is important to recognize with which complexity we wish to describe such spatiotemporal variable transport. In case of the PST model, we are satisfied if we know how the transport is for the entire area (for instance field) on average, and we are less interested in which part of the field the transport is fast or slow. In that case, the configuration of the solute front in x and y is of less importance, and we consider the field average front. For that front, we can be interested in the mean front position, its distribution, and its variance.

The field average front can be easily determined using Monte Carlo simulation, where random numbers are drawn for the distributed parameters, combined in the convective model, to determine the distributed front position at designated time. However, it is also possible to determine the field average front analytically, in case that the parameters v and R are lognormally distributed. This analytical procedure is as follows, for the simple case of a Heaviside solute input at the soil surface.

First, we experimentally or theoretically determine the mean (m) and standard deviation (s) of the lognormally transformed v and R, i.e. of $\ln(v)$ and $\ln(R)$. The so-called reproductive properties of the lognormal distribution are for the purely convective transport model of use to determine the statistics of front depth: if $z=vt/R$, then:

$$m_{\ln z} = m_{\ln v} + \ln(t) - m_{\ln R}$$

$$s_{\ln Z}^2 = s_{\ln v}^2 + s_{\ln R}^2 \tag{119}$$

As for the Heaviside input, the concentration at the soil surface changes abruptly from one value to another, we obtain different results depending on the initial and boundary conditions. As a first step, we determine, for a distributed $\ln(z)$ (for front depth), what is the probability that the front has not passed a certain depth. This probability is given by:

$$Pr\{z < z^*\} = \int_{-\infty}^{z^*} f_z dz \tag{120}$$

Inserting the lognormal pdf, with statistics as just determined, we obtain:

$$Pr\{z < z^*\} = \frac{1}{2} erf \left\{ \frac{\ln(z^*) - m_{\ln z}}{s_{\ln z}\sqrt{2}} \right\} \tag{121}$$

This probability is the probability that the front has not passed the depth z^*. To interpret that as a concentration, we have to reason what that means. If the front has not passed a certain depth, at that depth we still find the original (initial) concentration. Hence, Pr represents the fraction of the area, where we still find the initial concentration at depth $z=z_*$. We now consider two simplified systems. In the first case, the initial concentration is 1 and the incoming concentration is 0. In that case, the areally averaged concentration at depth z_* is:

$$\langle C(z^*) \rangle = Pr\{z < z^*\} \tag{122}$$

If, on the other hand, the initial concentration is equal to 0 and the incoming concentration is 1, then:

$$\langle C(z^*) \rangle = 1 - Pr\{z < z^*\} = \frac{1}{2} erfc \left\{ \frac{\ln(z^*) - m_{\ln z}}{s_{\ln z}\sqrt{2}} \right\} \tag{123}$$

More generally, if the initial and final (or input) concentrations are not equal to 0 or 1, the mean concentration can be transformed in the real value, using transformations such as:

$$\langle C \rangle = \frac{C(z^*) - C_{initial}}{C_{final} - C_{initial}}$$

(124)

Due to the logarithmic transforms, the mean concentration profile commonly looks as in figure below.

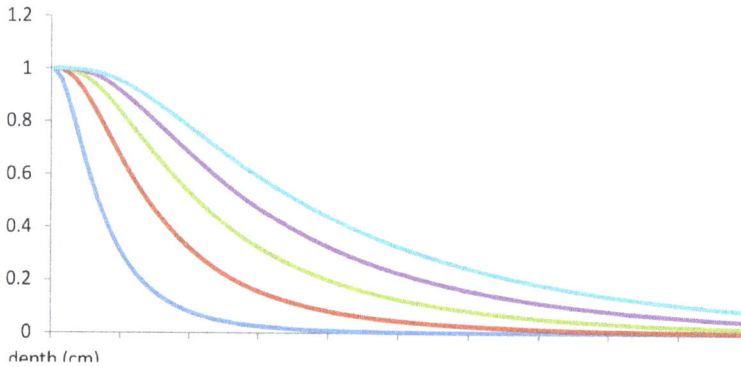

Areally averaged concentration (vertical axis) as a function of depth (horizontal), for different times: initial concentration 0 and final concentration 1.

In the presentation so far, we assumed that the important properties are lognormally distributed, i.e., taking the logarithmic transform, we obtain numbers that are normally distributed. Often, this assumption is appropriate, but also convenient, if the convective PST model is adopted: the reproductive properties hold for multiplication and division in case of a lognormal pdf, whereas they are additive/subtractive for the normal pdf. Also, lognormally distributed parameters cannot be negative, whereas normally distributed parameters can. This is important, because for instance the hydraulic conductivity and the retardation factor cannot be negative.

Although transport in the water unsaturated soil is predominantly vertically downward or upward, some horizontal displacement always happens. This transversal transport can also be important, for instance, if a slug of contaminant moves downwards, but dispersional mixing and movement in the transversal directions occurs for a contamination event of limited areal extent (in the horizontal x and y directions). In that case, the explained theory is valid but spatial moments have to be calculated in the other directions than only depths.

Air Flow through Soil

Various gases are found in the air space that exists between soil components. These include carbon dioxide, nitrogen, oxygen, etc. This chapter closely examines the air flow through soil and includes various topics such as soil gas, soil aeration, air permeability, etc.

Soil Gas

Soil Gas or Soil Air is a continuation of the atmospheric air. Unlike the other components, it is constant state of motion from the soil pores into the atmosphere and from the atmosphere into the pore space. This constant movement or circulation of air in the soil mass resulting in the renewal of its component gases is known as soil aeration.

Composition of Soil Air: The soil air contains a number of gases of which nitrogen, oxygen, carbon dioxide and water vapour are the most important. Soil air constantly moves from the soil pores into the atmosphere and from the atmosphere into the pore space. Soil air and atmospheric air differ in the compositions. Soil air contains a much greater proportion of carbon dioxide and a lesser amount of oxygen than atmospheric air. At the same time, soil air contains a far great amount of water vapour than atmospheric air. The amount of nitrogen in soil air is almost the same as in the atmosphere.

Composition of Soil and Atmospheric Air

	Percentage by volume		
	Nitrogen	Oxygen	Carbon dioxide
Soil Air	79.2	20.6	0.3
Atmospheric Air	79.9	20.97	0.03

Oxygen

The quantity of oxygen in soil air is less than that in atmospheric air. Plant roots and various microorganism require oxygen which they take from the soil air, thus, depleting the concentration of oxygen in the soil air.

The amount of oxygen also depends upon the soil depth. The oxygen content of the air in lower layer is usually less than that of the surface soil. This is possibly due to the more readily diffusion of oxygen from the atmosphere into the surface soil than in the sub-soil.

The quantity of oxygen is usually higher in dry season than during the monsoon. Because soils are normally drier during the summer months, opportunity for gaseous exchange is greater during this period. This results in relatively high O_2 and low CO_2 levels. Light texture soil i.e., sandy soil contains much higher oxygen percentage than heavy soil.

Carbon Dioxide

Decomposition of organic matter produces CO_2. Hence, soils rich in organic matter contain higher percentage of carbon dioxide. Production of CO_2 is associated with microbial activity, CO_2 increases with the increasing number and activity of microorganism.

Temperature and season also influence the carbon dioxide content in the soil air. High temperature during summer season encourages microorganism activity which results in higher production of CO_2. Soils on which crops are grown contain more CO_2 than fallow lands. The amount of CO_2 is usually much greater near the roots of plants than further away. It may be due to respiration by roots. The concentration of CO_2 is usually greater in sub-soil probably due to more sluggish aeration in lower layer than in the surface soil.

Water Vapor

Soil air, however, contains much more water vapor than atmospheric air. Capillary water in the soil is used to saturate the soil air with vapor. If the soil moisture falls below hygroscopic coefficient, the water vapor begins to decrease. During crop growing period, when soil remains moist, the amount of water vapor in the soil air would be more.

Soil Air in Relation to Soil and the Crop Management

Soil Management

The maintenance of a stable soil structure is an important means of augmenting good aeration. Pores of large size are soon drained of water following a rain, thereby allowing gases to move into the soil from the atmosphere. Maintenance of organic matter by addition of Farm yard Manure and crop residues and by growth of legumes is perhaps the most practical means of encouraging aggregate. Stability, which, in turn, encourages good drainage and better aeration.

In heavy textured (clay) soil, it is very difficult to maintain optimum aeration. Aeration in the soil can be added by controlling weeds and tilling the heavy soil. Consequently, no tillage (zero tillage) or minimum tillage practices, which the quite satisfactory on well-drained soils, have limitations on poorly drained soils.

Crop Management

Selection of crop is important criteria for adaption of crop in the soil. Alfalfa, fruits and forest trees and other deep-rooted plants require deep, well-aerated soils, such plants are sensitive to a deficiency of oxygen, even in the lower soil horizon.

In contrast, shallow-rooted plants, such as grasses, clovers etc. do well on soils that tend to be poorly aerated, especially in the subsoil. The rice plant flourishes even when the soil is submerged in water.

Soil Gas Measurements

Soil gas measurements are suited to identifying contaminants and sources. Soil gas surveys provide a screening method for detecting volatile compounds so that subsequent investigative activities can be located with the highest probability of contaminant detection. Technology used for soil gas measurements is most effective in detecting compounds with low molecular weights, high vapor pressure, and low aqueous solubilities. Soil gas surveys are often used to locate nonaqueous phase liquids (NAPLs), which are organic liquids or wastes that are sufficiently immiscible in water such that they may persist as a separate phase in the subsurface for many years.

There are multiple techniques available for soil gas analysis: active, passive, and flux chamber. An active soil gas survey consists of the withdrawal of a soil gas sample, typically from a perforated sampling probe, followed by analysis in a stationary or mobile laboratory. Passive soil gas surveys provide for the burial of probes containing absorbent materials to identify a broad range of volatile and semi-volatile organic compounds. Flux chamber measurements are primarily used in research applications, risk assessments, and toxicological studies when direct vapor flux values are desired. Soil gastesting technology is constantly evolving. Because identification is the major objective of soil gas analysis, it is important that the technology used to identify compounds is appropriate. Typical analytical issues include selectivity, detection limits, background levels, and cross-contamination.

If soil gas is used to determine concentration or mass, the ability of the sampling technology to quantitatively remove the soil gas of interest from a specified volume should be well understood. This is difficult to determine due to the heterogeneity of the soil matrix. In addition, it is difficult to determine the volume of soil gas removed. The use of soil gas to determine the total mass of an analyte should be undertaken with extreme caution. However, Hewitt has shown that using a robust and controlled vapor collection procedure may produce trichloroethene soil gas measurements that are quantitatively reliable. The radial area that is represented in a vapor sample will depend upon pump rate, pump time, hydrogeological factors and temporal effects such as temperature changes, precipitation, and activities within any overlying structure.

Soil Gas Sampling

Active soil-gas collection methods (as opposed to passive soil-gas sampling) involve, a vapor sample from a temporary or permanent probe inserted in the soil into a collection or analytical device. Samples are then transported to a laboratory, or in some cases they are analyzed on site.

Exterior soil-gas sampling is a screening tool used to rapidly and cost-effectively identify and delineate certain volatile contaminants in the subsurface. It is often used to ascertain the source, extent, and movement of pollutants, but it is not a substitute for groundwater sampling.

Sub-slab soil-gas sampling, where samples are taken though holes in buildings concrete slabs, is used to determine the potential for vapor intrusion. In some cases sub-slab samples are taken to determine whether to test indoor air, while in others it is conducted concurrently with indoor-air sampling to evaluate whether the subsurface is the source of indoor contaminants.

There are several types of soil-gas sampling instruments, including:

- Air-tight syringe: This is used to pull vapors from the soil matrix. The syringe is used to withdraw a soil-gas sample from a probe and inject it directly into an analytical instrument for on-site analysis. Some regulatory agencies require that samples collected by this method be analyzed within a few hours.

- Tedlar- bag: This requires a pump to pull the vapors from the soil. Regulatory agencies may require that samples collected by this method be analyzed within 28 to 48 hours. Bagged samples can also be drawn into sorbent tubes, which in turn undergo laboratory analysis.

- Glass bulb (or tube)- This has openings at each end, with one end having a valve where samples can be withdrawn with a syringe. The vapor sample is collected by connecting one end of the bulb to the probe and the other to a pump. The advantage of glass bulbs is that the glass is inert; also they are easy to use. The limitations of the glass bulbs are that they break easily and can leak contaminants through the valves. Sample holding times for glass bulbs are usually no more than 24 hours.

- SUMMA-canister: Under a vacuum, the canister pulls vapors out of the soil when the valve is opened. Each canister has a flow regulator and vacuum gauge between the vapor well and the canister. Regulatory agencies may require that samples collected by this method samples be analyzed within 30 days. The Summa canisters used for soil-gas sampling have a one to six liter sample capacity. (It is suggested that a one-liter canister be used if the sample is less than 5 feet below ground surface.)

- Hand-held direct measurement: This includes the Photo Ionization Detector (PID). PIDs provide immediate results, but they do not differentiate among contaminants or detect low levels of contaminants. More precise hand-held devices are under development.

- Flux chamber: This consists of an enclosed chamber that is placed on the surface for a specific period of time. This method yields both concentration data in the chamber and flux data (mass/area-time). Flux chambers are the least common soil-vapor survey method, and they are typically used to measure direct vapor migration from the subsurface to the surface.

Tedlar bags and Summa canisters need to be attached to a soil-gas wells or port. To install a soil-gas port, a hole is driven into the ground to a depth of four to five feet and a stainless steel or other non-reactive steel probe is inserted into the hole. The hole is then sealed around the top of the probe using modeling clay. (Some people recommend sealing the probe above the sampling zone with bentonite slurry for a minimum distance of 3 feet.) Soil-gas wells may be installed using a variety of drilling methods, such as direct push methods or augers.

Soil-gas samples should be sufficiently deep to minimize the effects of changes in barometric pressure (either ambient or within an overlying building due to the operation of heating, ventilation, and air conditioning systems), temperature, or breakthrough of ambient air from the surface.

The procedures for collecting sub-slab soil-gas samples are the same as for collecting sub-surface soil-gas samples, except that a slower flow rate and lower vacuum should be utilized to prevent ambient air from being drawn into the samples. Soil-gas sample collection techniques for vapor intrusion applications require much greater care than techniques historically used for typical site assessment applications because risk-based concentration levels for vapor intrusion scenarios are very low.

For a typical single family residential dwelling (approximately 1500 ft²), one vapor probe installed near the center of the slab is typically used to document the chemical composition of the sub-slab soil gas. Significantly larger dwellings (or other unique conditions in the subfloor or construction of the foundation) may require additional vapor probes.

Limitations and Concerns

Prior to selecting sample locations, locations of underground utility corridors should be well understood.

Heterogeneous soil conditions across a site under investigation can lead to poor delineation and misinterpretation of site contaminants. Data from areas of horizontal low permeability zones within the vadose zone may be misinterpreted as being an area of low contamination, and data from an area of high permeability in an otherwise low permeability area may be misinterpreted as an area of high contamination. High porosity areas such as sewer and utility trenches can serve as conduits for rapid vapor or gas migration, elevating gas readings some distance from the contaminated groundwater.

Recent research suggests that there may be substantial variation in soil gas concentrations beneath the slab. In some cases, where pressure differentials beneath the subsurface and indoors fluctuate between positive and negative, some soil gas contamination under the slab may be attributable to indoor contamination.

In vapor intrusion investigations, sub-slab soil-gas sampling is generally preferred over near-slab soil-gas sampling (within 10 feet horizontally of a building's foundation), and in general exterior soil-gas sampling (beyond 10 feet from the building footprint) is not acceptable as the primary line of evidence in the assessment of the vapor intrusion pathway.

Soil vapor samples collected under high vacuum conditions or under a continuous vacuum may contain contaminants that are "desorbed" and removed from the sorbed soil matrix and pulled from the dissolved phase into soil gas, rather than contaminants present in the undisturbed soil vapor. For collection systems employing vacuum pumps, the vacuum applied to the probe should be kept to a minimum necessary to collect the sample. To minimize the potential desorption of contaminants from the soil, Summa canisters should be filled at a rate less than half a liter per minute.

Also, because Summa canisters generally are under high vacuum, extra care should be exercised during sample collection to ensure that air from the surface is not being inadvertently sampled. The possibility of breakthrough from the surface increases as samples are collected closer to the surface (less than 5 feet below grade). To minimize the potential of surface breakthrough, the probe rod should be sealed at the surface.

Samples should not be exposed to light or extreme temperatures. Syringes and glass bulbs should be kept in cool dark locations. Samples should be covered or wrapped with foil and placed into an insulated container (cool but without ice).

The use of gas-tight glass syringes with Teflon˜ seals is preferred. The use of plastic syringes is discouraged because of the interaction between the plastic (or rubber) of syringes and some target analytes.

Soil-gas samples in Tedlar bags should not be subjected to changes in ambient pressure, because that could adversely affect the integrity of the bags. Increases in pressure may collapse the bag and decreases in pressure may expand the bag.

In general, soil-gas sampling events should be avoided after sizeable rainfall or even irrigation, because sampling in moist soil is unreliable. If no-flow or low-flow conditions are caused by wet soils or if water is drawn into a probe, sampling should be halted.

Other weather conditions may also hamper soil-vapor sampling. For example, condensation in the sample tubing may be encountered during winter sampling due to low outdoor air temperatures. Devices such as tube warmers may be used to address these conditions.

Using sub-slab soil gas sampling is questionable where the water table is high—that is, near (less than 2 feet below) the base of the sub-floor. Typically, vapors migrate through the most coarse and/or driest material. Depending on the analytical method, high moisture content in the soil-gas sample can "mask" results. Additionally, reduced permeability of the soil in the capillary fringe area may limit the movement of soil gas.

A common problem with this sampling method is soil probe clogging. A clogged probe can be identified by using an in-line vacuum gauge or by listening for the sound of the pump laboring.

In vapor intrusion evaluations, soil-gas sampling depth should be dependent on the depth of the contaminants as well as the vertical profile of the buildings, including basements and elevator shafts. Soil-gas samples should be obtained at appropriate depths so that the risk of human exposure can be adequately estimated. Some studies suggest that soil-gas samples collected at depths of 10 to 15 feet are a better indicator of vapor intrusion potential than samples collected at 5 feet where the source depth is greater than 15 feet beneath ground surface.

Exterior and near slab soil-gas samples should be collected at a minimum depth of 5 feet below the ground surface. In situations where the ground water table is less than 5 feet, alternative sampling protocols may have to be employed.

The results of the soil gas samples should not be averaged.

Residents are often reluctant to allow someone to drill a hole in their slab, especially if it's covered by a finished floor. To avoid less accurate near-slab sampling, investigators often seek to drill holes in closets or under carpets.

Applicability

Soil-gas sampling methods are used to screen of soil gas for Volatile Organic Compounds (VOCs) and Semi-Volatile Organic Compounds (SVOCs), as well as for gaseous mercury (metals) and radon (radionuclides). Data from exterior soil-gas surveys can be used to establish the extent of contamination at a site and to guide well placement and soil-boring programs. Soil-gas sampling is a commonly used to identify and evaluate contaminant movement and species. Sub-slab soil-gas sampling is a standard technique used to investigate potential vapor intrusion.

Factors Affecting Soil Air Composition

The composition of soil air is influenced by a number of factors such as nature of soil, soil condition, type of crop, microbial activity, season etc.

1. Nature of the Soil: Sandy Soils have macropore, as a result of which, aeration is very good in that soil. The soils that are water-logged contain small amount of oxygen as the pore space is filled with water immediately after a heavy rains or irrigation. The surface soil contains more macropore than the sub-soil. As a result, gaseous exchange is found to be more in surface soil than the sub soil. The oxygen percentage of Soil air varies with the depth of the soil and this is true in case of carbon dioxide also.

2. Soil organic matter: Soil organic matter is decomposed by microorganism present in the soil. Microbiological decomposition leads to the production of carbon dioxide and its content increases in the soil air. Hence soil rich in organic matter contains higher percentage of carbon dioxide.

3. Season: Season and temperature also influences the carbon dioxide content of the soil. The activity of soil micro-organism increases at high temperature during summer month which results in higher production of carbon dioxide. The composition of soil air shows marked seasonal variation, the intensity of which is affected by the texture of the soil and position of water table.

4. Soil moisture: The oxygen content of a soil decreases when the macropores are filled with water. But when the soil is artificially drained again, the macropores are filled with air and the oxygen content of soil increases.

5. Vegetation: Soils on which crops are grown contain more carbon dioxide than fallow land as a result of respiration of plant roots. The plant takes the soil oxygen and releases carbon dioxide. As a result, the carbon dioxide content of the cropped land increases near the root zone of the plant.

Soil Aeration

Soil aeration is phenomenon of rapid exchange of oxygen and carbon dioxide between the soil pore space and the atmosphere, in order to prevent the deficiency of oxygen and/or toxicity of carbon dioxide in the soil air. The well aerated soil contains enough oxygen for respiration of roots and aerobic microbes and for oxidation reaction to proceed at optimum rate.

Causes of Poor Aeration

Compact soils of finer textures suffer from poor aeration. Cultivation (working-the soil when crops are growing) of soil prevents it. Water logging is another important cause of poor aeration especially in the case of soils of finer texture.

The gaseous exchange between the soil air and the atmosphere may not be rapid enough to remove carbon dioxide from the soil air and to supply oxygen to the growing roots. This may happen if an excessive amount of readily decomposable organic matter has been added to the soil.

Mechanism of Gaseous Exchange

The exchange of gases between the soil air and the atmosphere takes place mainly by the following two mechanisms:

(i) Mass flow

Gases may move in a mass in the soil or out of it. The soil temperature is higher than the atmospheric temperature at midday when the soil gases expand and move out of the soil pore space into the atmosphere. The soil is cooler than the atmosphere during right when the atmosphere. The soil is cooler than the atmosphere during night when the atmospheric gases enter the soil.

When the atmospheric pressure is increased, volumes of gases present in the soil are decreased and therefore atmospheric gases enter the soil. Rain water displaces soil gases in the pore space and also carries gases dissolved in it to the soil. Rain fall usually account for 1/12 to 1/16th of the normal soil aeration. Variations in temperature and pressure between the soil and at the atmosphere play an insignificant role in soil aeration.

(ii) Diffusion

Most of the gaseous interchange between the soil and the atmosphere takes place by diffusion. Diffusion is the process by which each gas tends to move in the space occupied by another as determined by the partial pressure of each gas.

The partial pressure of a gas is the pressure which the gas would exert if it were present alone in the volume which has been occupied by the mixture of gases. The atmosphere contains higher amounts of oxygen than the pore spaces of soils which contain more carbon dioxide than the atmosphere.

So the partial pressure of oxygen is higher in the atmosphere than in the soil pore space and the partial pressure of carbon dioxide is higher in soil pore spaces than in the atmosphere even though the total pressure in the atmosphere and the soil pore spaces may be the same. So oxygen moves in the soil and carbon dioxide moves out of the soil.

Aeration Status of Soils

It can be determined in the three ways i. e:

1. Percentage oxygen and carbon dioxide content of the soil,

2. Oxygen diffusion rate, and

3. The oxidation reduction potential (Redox potential).

(i) Percentage Composition of Soil Air

The average inorganic soil contains about eight times more carbon dioxide and a little less oxygen than the atmosphere as shown below:

Table: Average composition of soil air an atmosphere

		(% by Vol)
	Atmosphere	Soil
Nitrogen	79.00	79.15
Oxygen	20.97	20.60
Carbon dioxide	0.03	0.25

Well aggregated soils contain enough macrospores to keep the soil aerated for proper growth and functioning of roots and micro-organisms. After a heavy rain, the macrospores are filled up with water but the soil may still contain some quantity of air dissolved in water. So micro-organisms can grow for a short time only, after which the soil must be drained so that the macrospores are re-filled with air.

(ii) Oxygen Diffusion Rate (ODR)

It determines the rate at which the oxygen should be supplied to the soil when it is being continuously used for the respiration of roots and soil micro-organisms. The growth of roots of most crops ceases when the oxygen diffusion rate decreases to about 20×10^{-8} gm./sq. cm/min. It should be above 40×10^{-8} gm./sq.cm/min. for good growth of most crops.

(iii) Oxidation-reduction Potential of Soil

The oxidation potential of chemical systems including soil is a measure of the tendency of the oxidation reaction to occur in that system, including soils.

This means that the system is in a reduced condition. Highly reduced soils have a high oxidation potential of +0.50 volts.

The reduction potential of a chemical system including soil is a measure of the tendency of the reduction reaction to occur in that system. This means that the system is an oxidized condition. Reduction Potential or Redox Potential (Eh) is the opposite in sign to the oxidation potential.

So a highly reduced soil which has a tendency to be oxidized has a Reduction Potential or Redox Potential Eh of- 0.50 volts. Well drained and aerated soils which are highly oxidized usually have a redox potential of +0.50 volts. The value of the Redox Potential increases when the oxygen content of soils decreases.

Factors Affecting Soil Aeration

(i) Soil Organic Matter

When organic matter is added to the soil, it is readily decomposed by the soil micro-organisms to liberate the carbon dioxide content of the soil air.

(ii) Since the top soil contains much more macrospore space than the subsoil, the opportunity for gaseous exchange is more in top soil than in the sub soil. Hence the oxygen content of the top soil is greater than that of the sub soil.

(iii) Soil moisture

The macrospores are filled up with water immediately after heavy rain when the oxygen content falls to near zero. When the soil is artificially drained again, the macrospores are filled up with air and the oxygen content of the soil increases.

Importance of Soil Aeration

Soil aeration affects the availability of some nutrients elements to plant roots. Manganese and iron occurs in the well aerated soil in their higher valent forms (Mn^{++++}, Mn^{+++}, Fe^{+++}) and in poorly aerated soils in their lower valent forms (Mn^{++}, Fe^{++}). They are available to plants only in their lower valent forms.

Crops suffer from manganese toxicity if an excessive amount of manganese occurs in the soil in the soluble form. Manganese toxicity to plant roots, under this circumstance, may be corrected either by making the soil more aerated by tilling the soil and improving drainage of the soil or by increasing the soil pH by applying lime to the soil.

When iron and manganese are in short supply to the soil, then the soil may be subjected to anaerobic condition by applying readily decomposable organic matter to it.

Carbon dioxide would be produced from the decomposition of organic matter to make the soil relatively more anaerobic when manganese and iron will be reduced from their respective higher Volant (Mn^{++++}, Mn^{+++}, Fe^{+++}) forms to their respective lower valent form (Mn^{++}, Fe^{++}) i.e. divalent forms and would be available to plant roots.

Ferric phosphate would be reduced to ferrous phosphate. Carbon dioxide produced from the decomposition of organic matter reacts with water to form carbonic acid which slowly dissolves insoluble phosphate. So the availability of phosphates (would be increased to the plant roots.

Sulphur occurs as sulphate in well aerated soil. Plant roots assimilate sulphate. Sulphate is reduced to sulphide in poorly aerated (water logged) soils. Hydrogen sulphide is toxic to plant a root which suffers from it in water logged soil.

Organic matter is decomposed by aerobic bacteria in well aerated soil when complex organic nitrogen and phosphorus compounds are decomposed to their respective simple inorganic compounds which plant roots readily assimilate symbiotic and non-symbiotic nitrogen fixation takes place only in well aerated soils.

Nitrates are reduced to oxides of nitrogen and nitrogen gases in poorly aerated soils. These gases escape to the atmosphere, long light coloured roots develop in well aerated soils. Root hairs develop best under well aerated condition.

Roots get thicker, shorter and darker in anaerobic soils that also retard the development of root hairs. Poor aeration causes abnormal development of roots, e.g. abnormal shaped sugar beet and carrot roots have been found in poorly aerated soils.

Nutrient absorption is an energy consuming process. Energy is available from respiration is expended in absorbing nutrient ions from the soil. Hence nutrient absorption is retarded in poorly aerated soils.

If an excessive amount of readily decomposable organic matter has been added to the soil, then it would decompose to evolve high amounts of carbon dioxide to the soil. Consequently root growth and germination of seeds would be adversely affected.

Some crops become infested with pathogens in poorly aerated soils. The incidence of will disease caused by the fungus (Fusarium sp) has been attributed to poor aeration. Citrus and suffers from die-back in poorly aerated soils have also reviewed the works of some investigators who have observed that poorly aerated soils (waterlogged soils) has an effect on the pathogenicity of root infesting fungi.

Soil Gas Movement in Unsaturated Systems

An understanding of gas transport in unsaturated media is important for evaluation of soil aeration or movement of O_2 from the atmosphere to the soil. Soil aeration is critical for plant root growth because roots generally cannot get enough O_2 from leaves.

Evaluation of gas movement is also important for estimating transport of volatile and semi volatile organic compounds from contaminated sites through the unsaturated zone to the groundwater. The use of soil venting, or soil vapor extraction, as a technique for remediating contaminated sites has resulted in increased interest in gas transport in the unsaturated zone. Migration of gases from landfills, such as methane formed by decomposition of organic material, is important in many areas. Soil gas composition has also been used as a tool for mineral and petroleum exploration and for mapping organic contaminant plumes. An understanding of gas transport is important for evaluating movement of volatile radionuclides, such as ^{3}H, ^{14}C and Rd from radioactive waste disposal facilities. The adverse health effects of radon and its decay products have led to evaluation of transport in soils and into buildings. A thorough understanding of gas transport is required to evaluate these issues.

Gas in the unsaturated zone is generally moist air, but it has higher CO_2 concentrations than atmospheric air because of plant root respiration and microbial degradation of organic compounds. Oxygen concentrations are generally inversely related to CO_2 concentrations because processes producing CO_2 generally deplete O_2 levels. Contaminated sites may have gas compositions that differ markedly from atmospheric air, depending on the type of contaminants.

Unsaturated media consist generally of at least three phases: solid, liquid and gas. In some cases, a separate nonaqueous liquid phase may exist if the system is contaminated by organic compounds. In most cases the pore space is only partly filled with gas. The volumetric gas content (θ_G) is defined as:

$$\theta_G = V_G/V_T$$

where V_G (L^3) is the volume of the gas and V^T (L^3) is the total volume of the sample. This definition is similar to that used for volumetric water content in unsaturated-zone hydrology. In many cases, the volumetric gas content is referred to as the gas porosity. The saturation with respect to the gas phase (S_G) is:

$$S_G = V_G/V_v$$

where V_v (L^3) is the volume of voids or pores. Saturation values range from 0 to 1. Volumetric gas content and gas saturation are related as follows:

$$\theta_G = \phi S_G$$

where ϕ is porosity (V_p/V_T). If only two fluids, gas and water, are in the system, the volumetric gas and water contents sum to the porosity. Therefore, volumetric gas content can be calculated if the volumetric water content and porosity are known.

In unsaturated systems, water is the wetting and gas is the nonwetting phase. Therefore, water wets the solids and is in direct contact with them, whereas gas is generally separated from the solid phase by the water phase. Water fills the smaller pores, whereas gas is restricted to the larger pores.

Mass Transfer Versus Mass Transport

Most of this chapter deals with transport of gas and associated chemicals through the unsaturated zone. Mass transfer refers to transfer of mass or partitioning of mass among gas, liquid and solid phases. Partitioning of chemicals into other phases retards their transport in the gas phase and can be described by:

$$C_T = \rho_b C_{ad} + \theta_l C_l + \theta_G C_G$$

where C_T is the total mass concentration ($M\,L^{-3}$ soil); C_G, C_l, and Cad are the mass concentrations in the gas and liquid phases and adsorbed on the solid phase; and r_b is the bulk density ($M\,L^{-3}$). The relationship between gas and water mass concentrations can be described by Henry's law:

$$C_G = K_H C_l$$

where K_H is the dimensionless Henry's law constant. Many types of isotherms describe the adsorption onto the solid phase, the simplest being the linear adsorption isotherm:

$$C_{ad} = K_d C_l$$

where K_d ($L^3\,M^{-1}$) is the distribution coefficient. The linear relationship generally applies to low polarity compounds.

If linear relationships are valid, total concentration can be written in terms of gas concentration as follows:

$$C_T = (\rho_b K_d / K_H + \theta_l / K_H + \theta_G) C_G = B_G C_G$$

where B_G is the bulk gas phase partition coefficient. In some studies this partitioning behavior is used to quantify the amount of a liquid phase in the system. For example, transport of gas tracers that partition into water or nonaqueous phase liquids is compared with transport of those that do not partition (conservative) into that phase to determine the amount of water or organic compound in the system.

Mechanisms of Gas Transport

Primary mechanisms of solute transport in the liquid phase include advection (movement with the bulk fluid) and hydrodynamic dispersion (mechanical dispersion and molecular diffusion). Mechanical dispersion, resulting from variations in fluid velocity at the pore scale, is the product of dispersivity and advective velocity. Transport in the gas phase may also be described by advection and dispersion. Although some studies have found mechanical dispersion or velocity-dependent dispersion to be important for chemical transport in the gas phase, in most cases mechanical dispersion is ignored because gas velocities are generally too small and the effects of diffusion are generally much greater than dispersion in the gas phase. Molecular diffusion coefficients are approximately four orders of magnitude greater in the gas than in the liquid phase.

Diffusive transport in the liquid phase is described by molecular diffusion. Traditionally, diffusive transport in the gas phase has also been described by molecular diffusion. Diffusion in the gas phase may, however, be much more complicated and may include Knudsen, molecular and non equimolar diffusion. Surface diffusion of adsorbed gases, generally not significant, is not discussed in this chapter. Pressure diffusion results from the separation of gases of different molecular weights under a pressure gradient and causes diffusion of heavier (lighter) molecules toward regions of higher (lower) pressure. Pressure diffusion is generally negligible at depths of less than 100 m, which include most unsaturated sections. Although temperature gradients in unsaturated media are generally too low to result in significant diffusion except at the land surface, thermal diffusion is important for water vapor transport.

Transport of a Homogeneous Gas

Transport of a homogeneous gas in dry, coarse-grained media can be described by advection. Because such a gas can be considered as a single component, the only type of diffusion possible is Knudsen diffusion. Single-component gases in dry, coarse-grained media are dominated by advective or viscous flux because Knudsen diffusion is negligible in such systems. This analysis of gas transport is appropriate when gas velocities are high. The single fluid, nonreactive, non-compositional approximation is appropriate only when one is interested in the bulk flow of a homogeneous gas.

Natural advective gas transport can occur in response to barometric pressure fluctuations, wind effects, water-table fluctuations, density effects, and can also be induced by injection or extraction, as in soil vapor extraction systems. Barometric pressure fluctuations consist of (1) diurnal fluctuations due to thermal and gravitational effects, which are on the order of a few millibars, and (2) longer term fluctuations that result from regional scale weather patterns, which are on the order of tens of millibars within a few hours of when a high or low pressure front moves through. The penetration depth of barometric pressure fluctuations increases with the thickness of the unsaturated zone and with the permeability of the medium. Because highs are balanced by lows, the net transport of the gas may be negligible, except in fractured media, where contaminants may migrate large distances. Smaller scale fluctuations, such as gusts and lulls related to wind, may be important in fractured media. Water-table fluctuations, resulting in changes in the gas volume, can produce advective flow; however, advective fluxes as a result of water-table fluctuations are considered important only if the rate of rise or decline of the water-table is rapid, the permeability of the material is high and the water table is shallow.

Gas Permeability

Gas permeability describes the ability of the unsaturated zone to conduct gas. Permeability (k) should be simply a function of the porous medium if the fluid does not react with the solid. Gas permeability $(k_G; L^2)$ is related to gas conductivity (K_G) as follows:

$$K_G = k_G \frac{\rho_G g}{\mu_G}$$

Some groups use the term gas permeability for K_G, which is called gas conductivity, and use the term intrinsic permeability for k_G, which is called gas permeability. Equation $(K_G = k_G \frac{\rho_G g}{\mu_G})$ shows that gas conductivity increases with gas density and decreases with gas viscosity.

The above material describes single-phase gas flow. In unsaturated media, the pore space generally contains at least two fluids, gas and liquid. Darcy's law is applied to each fluid in the system, which assumes that there is no interaction between the fluids. Because the cross-sectional area available for flow of each fluid is less than if the system were saturated with a single fluid, the permeability with respect to that fluid is also less. The gas permeability decreases as the gas saturation decreases. The relative permeability (k_{rG}) is a function of the gas saturation (S_G) and is defined as the permeability of the unsaturated medium at particular gas saturation $(k_G (S_G))$ divided by the permeability at 100% saturation (k_G):

$$k_{rG}(S_G) = k_G (S_G) / k_G$$

Relative permeability varies with (1) fluid saturation, (2) whether the fluid is wetting or nonwetting, and (3) whether the system is wetting or drying (hysteresis).

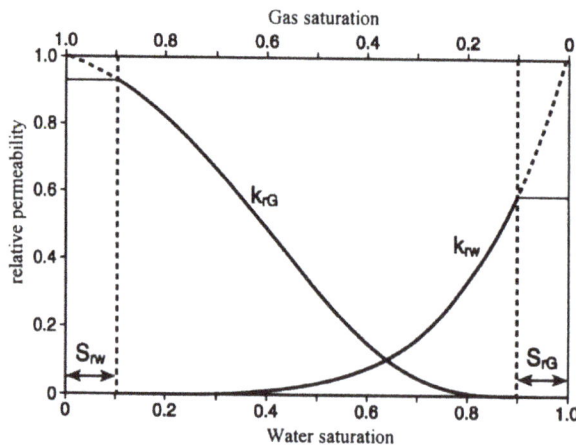

Schematic relative permeabilities with respect to gas and water saturation.

The relative permeability of the gas phase is greater than that of the liquid phase at low to moderate liquid saturations because the gas generally occurs in larger pores. Relative permeabilities do not sum to one. This has been attributed to flow pathways traversed by two fluid phases being more tortuous than those traversed by a single phase or to pores with static menisci that cannot result in flow. Zero relative permeabilities correspond to nonzero fluid saturations. For example, in systems initially water saturated that begin to drain, gas will not begin to flow until a minimum gas saturation has been reached, the residual gas saturation. The gas zero relative permeability region corresponds to trapped gas and disruption of gas connectivity caused by water blockages. Liquid-phase permeability also exhibits a zero-permeability region that corresponds to residual liquid or water saturation. In natural systems, the zone of residual water saturation corresponds to a zone of constant gas relative permeability because gas content does not increase. The dashed line in figure above in this zone corresponds with further increases in gas relative permeability (k_{rG}), which is related to increased gas content if the soil is oven dried and the water content is reduced to zero. A similar effect occurs with the water relative permeability (k_{rw}), where the dashed line shows increased water relative permeability corresponding to vacuum saturation or saturation of the sample under back pressure. Gas permeability is hysteretic at low water content, which indicates that gas permeability is not a unique function of gas saturation but depends on saturation history (i.e., whether the system is drying or wetting). At a given saturation, gas permeability is generally greater for wetting than for drying.

Various expressions have been developed to relate relative permeability to gas saturation. Many of the expressions were developed to estimate relative permeabilities with respect to water; the corresponding expressions for gas relative permeability were obtained by replacing the effective liquid saturation (S_e) by 1-S_e, where S_e is:

$$S_e = \frac{S - S_r}{1 - S_r}$$

where S_r is the residual water saturation.

Darcy's law can be written as:

$$J_G = -\frac{k_{rG}(S_G)k_G}{\mu_G}(\nabla P) = -\frac{k_{rG}(S_G)}{\mu_G}\nabla P$$

Expressions relating relative gas permeability and saturation. S_e is the effective saturation with respect to water; $S_e = (S_w - S_{rw})/(1 - S_{rw})$, S_{rw} is the residual wetting phase saturation (water in water-air system), λ is the bubbling pressure, and m and n are fitting parameters.

Brooks and Corey (1964)	$k_{rG} = (1 - S_e)^2 \left(1 - S_e^{\frac{2+\lambda}{\lambda}}\right)$	
Corey (1954)	$k_{rG} = (1 - S_e)^2 (1 - S_e^2)$	
Falta et al. (1989)	$k_{rG} = (1 - S_e)^3$	
van Genuchten (1980); Mualem (1976)	$k_{rG} = (1 - S_e^{0.5})\left\{1 - \left[1 - (1 - S_e)^{1/m}\right]^m\right\}^2$	$m = 1 - 2/n$ $0 < m < 1$
van Genuchten (1980); Burdine (1953)	$k_{rG} = (1 - S_e)^2 \left\{1 - \left[1 - (1 - S_e)^{1/m}\right]^m\right\}$	$m = 1 - 2/n$ $0 < m < 1; n > 2$

The velocity of the gas particles (V_G) $(L\,t^{-1})$ can be calculated from the volumetric flux density by dividing by the volumetric gas content:

$$V_G = J_G/\theta_G$$

Field Techniques

Estimation of Gas Permeability for Advective Gas Flow

Advective transport of gases depends on gas permeability and pressure gradient. Gas permeability can be estimated from (1) analysis of atmospheric pumping data, (2) pneumatic tests, or (3) measurements by air minipermeameters. Comparison of gas permeabilities from laboratory and field indicates that field derived estimates of gas permeabilities generally exceed laboratory derived estimates by as much as orders of magnitude. These differences in permeability are attributed primarily to the increase in scale from laboratory to field measurements and inclusion of macropores, fractures and heterogeneities in field measurements. Field permeability measurements in low permeability media include the effects of viscous slip and Knudsen diffusion.

Analysis of Atmospheric Pumping Data

Comparison of temporal variations in gas pressure, monitored at different depths in the unsaturated zone, with atmospheric pressure fluctuations at the surface can be used to determine minimum vertical air permeability between land surface and monitoring depth. Differential pressure transducers are used to monitor gas pressures in the unsaturated zone. Gas ports generally consist of screened intervals in boreholes of varying diameter. Flexible tubing (Cu or nylon) connects the gas port at depth with a differential pressure transducer at the surface.

One port of the transducer is left open to the atmosphere. Atmospheric pressure is monitored at the surface by a barometer.

Data analysis consists of expressing the variations in atmospheric pressure as time harmonic functions. The attenuation of surface waves at different depths in the unsaturated zone provides information on how well or poorly unsaturated sections are connected to the surface. The accuracy of the results increases with the amplitude of the surface signals. Pressure fluctuations resulting from irregular weather variations change by as much as 20 to 30 mbar (2,000 to 3,000 Pa) during a 24-h period.

If the surface pressure (upper boundary condition) is assumed to vary harmonically with time, and the water table or a low-permeability layer acts as a no-flow boundary, an analytical solution can be derived. Pneumatic diffusivity can be estimated graphically by means of the amplitude ratio. The ratio of the amplitude at a certain depth z compared with the amplitude at the surface can be obtained graphically or by time-series analysis. Air permeability is estimated from the pneumatic diffusivity by dividing by the volumetric air content.

Pneumatic Tests

Pneumatic tests are also used widely to evaluate gas permeability in the unsaturated zone. In these tests, air is either injected into or extracted from a well and pressure is monitored in gas ports installed at different depths in surrounding monitoring wells. A reversible air pump is used to inject or extract air. Most analyses of pneumatic tests assume that the gas content (θ_G) is constant over time, i.e., that no redistribution of water occurs during the test. To evaluate this assumption, injection or extraction tests should be conducted at several different rates. If results from the different rates are similar, the assumption of constant gas content is valid. The tests can be conducted in horizontal or vertical wells.

A variety of techniques are available for analyzing pneumatic tests. The initial transient phase of the test or the steady-state portion of the test can be analyzed. If transient data are available, volumetric gas content can also be estimated. Analysis of pneumatic tests resembles the inverse problem in well hydraulics, where permeabilities are estimated from pressure data. Solutions for estimating gas permeability differ in terms of the boundary conditions that are assumed at the ground surface (such as unconfined, leaky confined and confined) and the method of solution. The lower boundary is generally assumed to be the water table or an impermeable layer. All solutions assume radial flow to a vertical well.

The gas flow equation is nonlinear because of the pressure dependence of the density, viscosity and permeability (Klinkenberg effect). In many cases the pressure dependence of the density is approximated by ideal gas behavior (equation $(PV = nRT$, $\rho_G = \dfrac{nM}{V} = \dfrac{PM}{RT}$)). Under low to moderate pressures and pressure gradients typical of unsaturated media, pressure dependence of the viscosity can be neglected. In most analyses, the Klinkenberg effect is also neglected. If pressure variations are assumed to be small, the transient gas flow equation ($\dfrac{\theta_G \mu_G}{P_o} \dfrac{\partial P}{\partial t} = \nabla(k_G \nabla P)$) can be writ-

ten in a form similar to that of the groundwater flow equation ($S_s \dfrac{\partial h_w}{\partial t} = \nabla(K_w \nabla h_w)$):

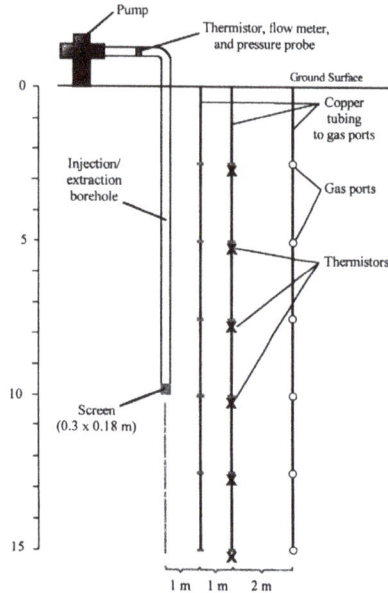

Schematic design for a field pneumatic test.

$$\frac{\theta_G \mu_G}{P_o} \frac{\partial P}{\partial t} = \nabla(k_G \nabla P)$$

$$S_s \frac{\partial h_w}{\partial t} = \nabla(K_w \nabla h_w)$$

where S_s is specific storage, h_w is hydraulic head (L) and K_w is hydraulic conductivity ($L\,t^{-1}$).

Applied Numerical Modeling

Single Phase Flow

A summary of the various types of codes available for simulating gas flow is provided in table. Most numerical simulations of water flow in unsaturated media have generally ignored the gas phase by assuming that the gas is at atmospheric pressure. The equation that is solved is the Richards equation, which describes a single-phase (liquid), single-component (water) system. The Richards approximation is generally valid for most cases of unsaturated flow.

A variety of numerical models have been developed to simulate gas flow in unsaturated systems. An important consideration in choosing a code for evaluating gas transport is the assumptions of each code. Groundwater flow models can be used to simulate gas flow in cases where the vapor behaves as an ideal gas and where pressure fluctuations are small and gas content is constant (no water redistribution). Such assumptions are generally valid for vapor extraction remediation systems. The code most widely used to simulate groundwater flow is MODFLOW. Joss and Baehr developed a sequence of computer codes (AIR3D) adapted from MODFLOW to simulate gas flow in the unsaturated zone. The codes can be used to simulate 3-D air flow in a heterogeneous, anisotropic system including induced air flow in dry wells or trenches. Pre- and postprocessors are included. AIR3D can also be used to simulate natural advective air transport in response to barometric pressure fluctuations in shallow, unsaturated systems when gravity and temperature gradients can be neglected. AIR3D transforms the air flow equation ($\frac{\theta_G \mu_G}{P_0} \frac{\partial P}{\partial t} = \nabla(k_G \nabla P)$) into a form similar to the groundwater flow equation ($S_s \frac{\partial h_w}{\partial t} = \nabla(K_w \nabla h_w)$) that is solved by MODFLOW. Air compressibility is approximated by the ideal gas law. The simulations can be conducted (1) in a calibration mode to evaluate parameters such as permeability from pneumatic tests or (2) in a predictive mode.

Table: Summary of types of codes available to simulate gas flow.

Code	Dim.	No. Phases	Components	Energy Balance	Porous/Fractured Systems
AIR3D	3	1	gas	no	porous
BREATH	1	2	water	yes	porous
SPLaSHWaTr	1	2	water	yes	porous
UNSAT-H	1	2	water	yes	porous
Princeton Code	2	2	water and air	no	porous
FEHM	3	2	water and air	yes	porous/fractured
TOUGH	3	2	water and air	yes	porous/fractured
STOMP	3	3	water, air, NAPL	yes	porous/fractured

Air Permeability

Air permeability of soil (and other porous materials) is the coefficient, k_a, governing convective transmission of air through soil under an applied total pressure gradient. The theory for the flow of air through soil is based on Darcy's law, which states that the velocity of a fluid flowing through a porous column is directly proportional to the pressure difference and inversely proportional to the length of the column. This has been investigated thoroughly for both isotropic and anisotropic media. Importantly, large pores and wide cracks contribute most to air permeability because the volumetric flow of air through a single pore varies as the fourth power of the pore radius. The air permeability coefficient, k_a, has units of m² and is also known as the intrinsic permeability to air. It can be derived from Darcy's law (for laminar flow of liquids) using simple assumptions about isothermal, non-turbulent flow of a viscous gas.

It has been used since the early part of the twentieth century for describing, defining, and characterizing the structural arrangement and continuity of the pore space in soils. Air permeability is very sensitive to differences in soil structure and has been widely used to characterize the changes in structure that result from different soil management practices. It has even been used to predict other important soil physical properties such as the saturated hydraulic conductivity.

Measurement of air permeability in the field (as opposed to in the laboratory) is a desirable technical feature but the surface layers of the soil are often structurally anisotropic, so variability is large and the interpretation of field data can be difficult.

Air permeability of soils can be measured in the laboratory in a number of ways, but two principal methods will be discussed here, namely the constant pressure gradient method (with a measured flux of air) and the constant flux method (with a measured pressure). The choice of method is usually a matter of convenience, but each method has its own advantages.

Constant Pressure Gradient Method

Numerous variations of the constant pressure gradient method have been used for measuring air permeability. The method consists of exposing a soil sample in a defined volume (e.g., cylindrical ring) to a large volume of air having constant pressure (above atmospheric), and measuring the volume of air that passes through the soil core with time. The intrinsic permeability to air, ka (m²), is calculated for small air pressures (i.e., $<0.2\,\mathrm{m}\,H_2O$):

$$\frac{Q}{A} = k_a \frac{\rho_w g}{\eta} \frac{\Delta h}{L}$$

where Q is the volume of air measured at the high-pressure side of the soil core (inlet) passing into the soil core per unit time (m³ s⁻¹), A the cross-sectional area of the soil core (m²) orthogonal to direction of airflow, L the length of the soil core (m), $\Delta h = h_i - h_a$ the difference in air pressure expressed (m) across the length of the soil sample between the air-inlet side, i, and the air-outlet side, a, at atmospheric pressure, $h_i = P_i / \rho_w g$ the inlet pressure head (m), $h_a = P_a / \rho_w g$ the outlet pressure head (m), ρ_w the density of water (kg m⁻³), g the gravitational acceleration constant (ms⁻²), and η the viscosity of air (kg m⁻¹ s⁻¹).

Materials

Schematic specifications for this apparatus can be found in Tanner and Wengel, but many variations are possible. Only the basics are given here, illustrated in figure below.

1. Air tank of large volume (i.e., at least 20 L or 200 times larger than the sample volume), made from either stainless steel or Nalgene, so long as it is rigid; the tank is partially filled with water to trap air under the float-can.

2. Float-can, which is made of either stainless steel or Nalgene.

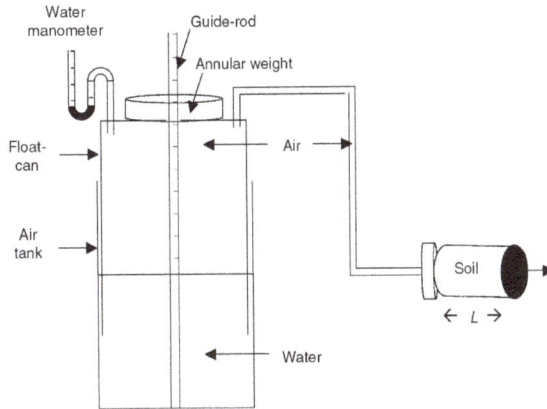

Constant pressure gradient apparatus to measure air permeability.

3. Guide-rod (calibrated in advance using water to produce a measure of m^3 air mm^{-1} on the guide-rod); this ensures the float-can sinks evenly into the water reservoir in the air tank and also serves as a convenient measuring stick to determine the volume of air flowing out of the air tank into the soil core.

4. Annular dead-weights: These may be required to increase the pressure in the air tank; these are fitted over the guide-rod and placed on the float-can.

5. Pressure gauge: A simple water manometer will suffice.

6. Soil sample: A soil core (at a specified soil water matric head) collected in a rigid cylindrical sample-holder is required.

7. Sample-holder: Depending upon resources, this can be either an elaborate device or a simple one. The simplest construction consists of a short, empty ring of the same (outside) diameter as the ring containing the soil core. A rubber bung and air-supply-tube are fitted to one end of the empty ring and the other end is joined to the soil core using a strip of Parafilm. (More elaborate designs can be conceived, of course, and we generally use a specially constructed brass ring, the diameter of which is just large enough to contain a rubber O-ring fitted snugly around the ring containing the soil core. This avoids the tedium of cutting and stretching multiple strips of Parafilm.)

Procedures

1. Cylinder containing the soil core is connected to the sample-holder using either pre-cut strips of Parafilm or by fitting it directly into a specially constructed brass ring with rubber O-ring to ensure a good seal. The sample-holder is connected to a laboratory retort stand and placed into a shallow bucket of water such that the connection between the cylinder and its sample-holder is immersed (to check for leaks), while the outflow side of the soil core is left open to the atmosphere.

2. Float-can is allowed to fall freely under its own weight, or with the addition of annular weights to generate higher air pressures maintained throughout the experiment.

3. Guide-rod is monitored with time and when the rate at which the float-can falls becomes constant (steady-state), two readings are selected to calculate the flux of air across the sample under the measured constant pressure difference.

4. Intrinsic permeability is calculated using equation ($\frac{Q}{A} = k_a \frac{\rho_w g}{\eta} \frac{\Delta h}{L}$).

Constant Flux Method

With the constant flux method, a constant flux of air is imposed across the soil sample and the resulting air pressure difference is measured. Methods to impose air fluxes are easily constructed and depend only on the resources at hand in the laboratory. A small cylinder and flow meter supplying compressed air, for example, would suffice. Alternatively, figure shows the use of a syringe whose air volume, V, is delivered by a slowly advancing, motordriven piston. Any other apparatus that can supply a nonvariable and easily measured air flux is adequate.

Steady-state flow of air may or may not occur, and when it does, the pressure difference can become quite large ($\Delta h > 0.2$ m H_2O). For large values of Δh, the intrinsic air permeability, k_a, should be calculated in accordance with the ideal gas law, as follows:

$$\frac{Q}{A} = k_a \frac{\rho_w g}{\eta} \frac{\Delta h}{L} \left[1 - \frac{\Delta h}{2h_i} \right]$$

where hi and the constants ρ_w, g, and η are defined above. If Δh does not exceed 0.2 m H_2O, the second term in parentheses approaches zero, so both terms in parentheses approach unity and may be ignored. In that case Equation ($\frac{Q}{A} = k_a \frac{\rho_w g}{\eta} \frac{\Delta h}{L} \left[1 - \frac{\Delta h}{2h_i} \right]$) reduces to Equation ($\frac{Q}{A} = k_a \frac{\rho_w g}{\eta} \frac{\Delta h}{L}$).

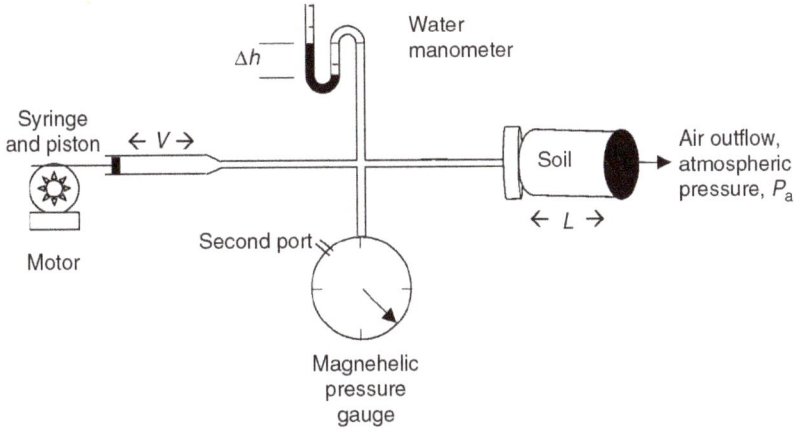

Constant flux apparatus to measure air permeability.

Materials

The apparatus is shown in figure above.

1. Syringe: This can be relatively large, having volume, V, anywhere from 35 to 140 cm³; the inner walls of the syringe can be greased with Vaseline to prevent air leakage. It is convenient to have a range of different sizes of syringes on hand so that the flux of air can be varied easily.

2. Motor-driven piston: The "piston" is simply the plunger for the syringe, which is placed on a motor-driven carriage and calibrated for each syringe and motor speed by recording the mass of water delivered per unit marked on the syringe, which delivers air at fluxes set according to the speed of the motor.

3. Pressure gauge: Either a simple water manometer or a Magnehelic pressure gauge is suitable. Conditions for use of each are outlined below.

4. Soil sample: Same as for the constant pressure gradient method discussed in Material of constant pressure gradient method.

5. Stiff connection tubing of small inside diameter to minimize the "dead air" volume.

Procedures

1. As for the constant pressure gradient method, the soil sample is connected to a rigid holder fitted with a rubber O-ring to ensure a complete seal. The sample-holder is connected to a laboratory retort stand and placed into a shallow bucket of water such that the connection between the cylinder and its sample-holder is immersed (to check for leaks), while the outflow side of the soil core is left open to the atmosphere.

2. Syringe is connected to the soil sample and the motor-driven "pump" is set to move at a constant velocity. As it advances, it builds up an air pressure difference across the soil core. This pressure difference is monitored using the Magnahelic pressure gauge shown in figure. The actual settings that you choose will depend upon how permeable the soil sample is. Some initial adjustments of the flux with different-sized syringes and pump-settings are often needed prior to data collection. When the pressure difference becomes constant, the mass-flux of air through the soil core also becomes constant.

3. When steady-state flow is established and both the flux and the pressure difference across the soil core are constant, the intrinsic permeability to air can be calculated using equation ($\frac{Q}{A} = k_a \frac{\rho_w g}{\eta} \frac{\Delta h}{L} \left[1 - \frac{\Delta h}{2h_i} \right]$).

4. If steady-state airflow conditions are not reached, the intrinsic permeability to air must be calculated from observations taken during the transient state. In this instance, a water manometer needs to be used rather than the Magnehelic pressure gauge because the Magnahelic pressure gauge "bleeds" air until steady-state conditions are reached.

References

- Soil-air-composition-and-relation-1133: soilmanagementindia.com, Retrieved 25 May 2018

- Scanlon-Soil-Phys-Comp-02: beg.utexas.edu, Retrieved 25 April 2018

- Soil-aeration-definition-factors-and-importance-3509: soilmanagementindia.com, Retrieved 19 June 2018

- Soil-gas: agricultural-and-biological-sciences: sciencedirect.com, Retrieved 14 March 2018

- Soil-air-composition-and-factors-soil-science-15767: soilmanagementindia.com, Retrieved 05 March 2018

Permissions

Index

www.ingramcontent.com/pod-product-compliance
Lightning Source LLC
Chambersburg PA
CBHW061957190326
41458CB00009B/2898